4 Chemistry of Plant Protection

K. Naumann

Synthetic Pyrethroid Insecticides: Structures and Properties

With 35 Figures and 114 Tables

Springer-Verlag Berlin Heidelberg GmbH

Author:

Dr. Klaus Naumann
Bayer AG, Pflanzenschutzzentrum Monheim
Gebäude 6550
5090 Leverkusen, Bayerwerk/FRG

Managing Editors:

Dr. G. Haug
Pflanzenschutzzentrum Monheim, Bayerwerk
D-5090 Leverkusen/FRG

Prof. Dr. H. Hoffmann
Pflanzenschutzzentrum Monheim, Bayerwerk
D-5090 Leverkusen/FRG

This series continues the handbook
"Chemie der Pflanzenschutz- und Schädlingsbekämpfungsmittel"
edited by R. Wegler

ISBN 978-3-642-74851-6   ISBN 978-3-642-74849-3 (eBook)
DOI 10.1007/978-3-642-74849-3

Library of Congress Cataloging-in-Publication Data
Naumann, K. (Klaus), 1939 — Synthetic pyrethroid insecticides: structures and
properties/K. Naumann.
p. cm. — (Chemistry of plant protection: 4)
Includes bibliographical references.

1. Pyrethroids — Synthesis. 2. Insecticides — Patents. I. Title. II. Series.
SB952.P88N38 1990 668'.651 — dc20 89-28247 CIP

2152/3020-543210

Editorial Board

Editorial

The series "Chemistry of Plant Protection" is the successor to our handbook "Chemie der Pflanzenschutz- und Schädlingsbekämpfungsmittel" which was edited by R. Wegler.

Aims and Scope

Plant protection chemistry continues to develop rapidly. Important targets have been achieved and improvements with respect to selectivity, efficacy, mode of action, environmental and toxicological compatibility have been made. There have been new developments in the field of formulation and side-effect research. New classes of substances with high effectivity at very low dosages have been discovered.

These developments have been accelerated by new biological and biochemical discoveries as well as by advances in synthetic chemistry. In order to emphasize the interaction between these and other related disciplines, the formerly distinct presentation of different fields of crop protection chemistry, such as insecticide, fungicide or herbicide research, has been abandoned. The following volumes will contain recent developments in research on new active chemical substances as well as reports on metabolism, residue analysis, biochemical mechanisms, and other important innovations.

The volumes will be more or less topic-oriented. Each volume will have a "volume index" which approximately characterizes the contents. The aim of the publisher and the board of editors is to produce review articles of high quality by leading scientists in the field of plant protection. Suggestions for such contributions from all those involved in plant protection are, of course, always welcome.

Editorial Policy

The series publishes critical review articles in English from invited authors. The topics should be covered comprehensively and the international literature evaluated. Ideally, contributions should comprise 40–80 typewritten pages. Experimental details, except when not previously published, should be covered by citing appropriate reference. The responsible editor, that is the editor who invited the article, discusses the scope of the review with the author on the basis of a tentative outline which the author is asked to provide.

*Dedication
to my patient family*

Preface

The idea of Volums 4 and 5 of this series is a combination of a very condensed but broad review, handbook and textbook on most of the theoretical and practical aspects of pyrethroids of interest to chemists, biologists, pharmacologists, toxicologists and other people involved in insecticidal research, development, ecotoxicology, application, patenting and commercialization of pyrethroids, as seen from a point of view of an industrial chemist once actively in volved in pyrethroid research.

In order to provide a model data base for testing quantitative structure activity relationships, the unique wealth of the numerous and diverse biological data dispersed in the literature is listed in about 110 tables, put in relation to each other and other standard compounds, together with many structural formulas of the compounds involved. A number of data are published for the first time. Some of the more important recent QSAR-studies are briefly acknowledged. Conformational aspects of bioactive pyrethroids are discussed (from synthesis, QSAR, X-ray); some of the X-ray derived conformations are published for the first time.

The molecular basis of pyrethroid biological action is taken into consideration as intensively as possible. Stereochemical aspects are considered wherever they are involved in synthesis, biological activity, mode of action, metabolism (Vol. 4), and particularly in the complex problems in the production and relation of isomerically enriched trade products (Vol. 5).

Since pyrethroid research is now, in many aspects, a finished chapter of applied research, a historical treatment of the course of inventions is given (inventions, patent priorities).

A detailed, but condensed presentation of the numerous chemical synthetic aspects connected with pyrethroid chemistry is given in Vol. 5 with 270 schemes of chemical formulas, to show to the scientific public, particularly to students, the wealth of splendid chemistry performed mostly by industrial chemists.

The very comprehensive collection of patents (Vol. 5) for active ingredients serves two purposes:

a) rapid information on what has been published for patent professionals and industrial chemists,

b) a documentation for students and other researchers not involved in this kind of research. They can see, how research in
   a fruitful area develops over the years starting from an interesting lead compound.

A short overview on commercial and marketing matters of
pyrethroids in connection with the agricultural application rates
of commercial pyrethroids is included. All commercial and most
developmental pyrethroids are characterized by their various
names, code numbers, isomeric relations and commercial pro-
ducers (Vol. 4).

Alltogether about 2600 references from the scientific and patent
literature are evaluated.

Leverkusen, January 1990   Klaus Naumann

Table of Contents

Acknowledgement

I have to thank my collegue Dr. Fuchs for a critical checking of the chemical aspects of this volume, Dr. Leicht for looking over the biochemical and physiological topics, Mrs. Kuenhenn and particularly Mrs. Logemann for preparing the type written manuscript. I have to thank also Dr. Rabe for a very carefull reading word by word in order to pick out occasional assaults on the English language and grammar and some other errors. Mrs. Fischer and Mr. Wiemann produced the mumerous chemical formulae by means of the computer graphics program FDS Gral. I am particulary gratefull to Dr. Born for providing some X-ray analysis data prior to his own publication. I appreciate the contribution of Dr. Zywietz in producing the stereoformulae of some toxins and of the X-ray structures of six pyrethroids. Mr. Frössler drew the figures. Finally I have to thank Prof. Hoffmann and the Bayer AG for providing resources and experts of the company, to turn the original manuscript into a state suitable for print. During this process I appreciated the encouraging comments of P. Enders of Springer-Verlag, in order to save as much as possible from the original draft for the final printed version.

A. Biologically Active Pyrethroids

1 Naturally Occuring Pyrethrins

The pyrethrins are constituents of pyrethrum, pelitre or buhach, obtained from the pretty daisy-like flowers *Chrysanthemum cinerafolis (cinerariae folium = Tanacetum cin.)* and *C. coccineum (= roseum = carenum)* which grow naturally on the hills of the Caucasus mountains and in Dalmatia. In comparison to most of the other known 'insecticidal' constituents of other plants, these compounds have a high and rapid action against a great number of insects, while being harmless for animal, man and the environment. Their very low stability under the atmospheric influences of oxygen and light however, precludes the use against agricultural pests.

The dried flowers were used as insecticides in ancient China and later on in the Middle Ages in Persia. They became known to Europeans via Armenian traders almost two hundred years ago as 'Persian dust' or 'insect powder'. As soon as their

Table 1. Ingredients of natural Pyrethrum

Acid component	Alcohol component		
	Pyrethrolon	Jasmolon	Cinerolon
Chrysanthemic acid	Pyrethrin I 35%	Jasmolin I 5%	Cinerin I 10%
Chrysanthemumdicarboxylic acid Pyrethric acid	Pyrethrin II 33%	Jasmolin II 4%	Cinerin II 14%

secret of origin became evident the commercial cultivation of *Chrysanthemum cinerafolis* in Dalmatia, much later in the last century in Japan, Brazil and USA, was the source of this insecticide. Nowadays large plantations and many small holdings in Kenya and Tanzania and some other places, like Colombia, provide this ingredient as an extract or dried flowers. The flowers in Kenya contain on an average 1.3% (2–3 mg, sometimes up to 4 mg per flower) pyrethrins. Shortly after flowering they are picked [1] and dried or extracted [2–4]. The powdered dried flowers are the original trade item. For extraction of the pyrethrins, combinations of methanol and kerosin [5–9] or petrolether/acetonitrile [9] or petrolether/nitromethane are suitable. The last mixture provides a 90% concentrate of the 6 pyrethrins (Table 1).

These are the optically active esters of chrysanthemic acid (monocarboxylic acid) and pyretric acid (dicarboxylic acid) [10, 11] and substituted optically active cyclopentenolons [11, 12]. Pyrethrin II is less active, but faster acting [13] than pyrethrin I. Pyrethrin I is the most active constituent of pyrethrum [14, 15]. Pyrethrin II contains 5 stereochemical discriminators:

1-R; 3-S; E; S'; Z'
($\cong$ 1-R trans E; S'; Z')

The absolute configuration of the acid part is 1-R-trans- [16, 17] and E; S of the alcohol, and Z in the side chain [18]. Pyrethrin I is about 100 times more active than the other constituents against the housefly.

Not much is known yet of the biosynthesis of pyrethrins. The alcohol of pyrethrin I and II is not just made of acetate-building blocks [19]. Precursors of chrysanthemic acid seem to be mevalonic acid [20, 21, 22] and chrysanthemol-pyrophosphate [23]. Chrysanthemol is found also in a different plant Artemisia ludoviciana [24]. Another constituent of pyrethrum is the linolic ester of pyrethrolon [25].

Due to its favorable properties for indoor application against insect pests on food or in the close vicinity of humans, pyrethrum is still a product in demand, which is only one of the reasons it is expensive. A few years ago, Kenia and Tanzania produced about 23,000 tons annually corresponding to 90% of the world harvest [26]. In the meantime this situation has changed, but not due only to the advent of even more active synthetic pyrethroids. Less favorable conditions for efficient production in these countries are also responsible. Since the pyrethrins are still of interest, other research revealed, that chrysanthemum farming is feasible even under normal agricultural conditions [27], and that pyrethrum producing cell cultures may become interesting competitive alternatives [28–33]. On the other hand the search for more efficient chrysanthemum varieties yielded not only strains with higher contents (average 1.6%), but also ones which are devoid of this interesting and expensive agricultural product, e.g. *Chrysanthemum balsamita* [35], *C. corymbosum*, *C. partemicum* and *C. indorum* [36].

Many more details on pyrethrum history, properties and applications are compiled in Casida's book "Pyrethrum – The Natural Insecticide" (1972) [37]. See also Otieno [38].

2 Synthetic Pyrethroids

2.1 General Remarks on the Structure and Activity of Synthetic Pyrethroids

What is a pyrethroid? This question is nowadays not so easy to answer as 20 years ago, when it was clear, that it meant synthetic insecticides with very lipophilic properties, whose structural relations to the insecticidal parent lead components of pyrethrum could easily be discerned.

In the past 16 years the structural variations have arrived at structural formulas which do not bear any formal resemblance for the average chemist any more when they are drawn in the plane. Nevertheless, by virtue of the very similar 3-dimensional shape of certain conformations and the very similar physical properties of the 'skin' of their common spacial volume (in terms of size, charge distribution, lipophilic properties, electrondonating and − accepting properties) and their identical or principally similar mode of action and target site at the nerve membrane respectively (despite different observable symptoms in vivo and vitro), they can be grouped together and distinguished from all the other insecticides, except DDT. This old insecticide belongs molecularpharmacologically and electrophysiologically to this group, but possesses only a part of the typical structural features which are displayed in abundance in the collection of structures in the following tables, the following structure-activity-chapter and in the collection of patent applications in the second volume of this book, the Vol. 5 of this series.

In short: A pyrethroid is lipophilic, looks similar to the structures on pages 7–9, has a strong effect on nerve membrane sodium channels, like the natural archetypical compound pyrethrin I, and kills insects.

Despite this pedigree the name pyrethroid, in French pyrethrinoid, for the sake of clearness, should not be mixed up in commercial products with the name pyrethrum. "Synthetic pyrethrum" or "-pyrethrin" means the nature-identical product.

Insecticidal Action and Structural Variability of Synthetic Pyrethroids in the History of Pyrethroid Research

Besides a whole set of structural demands, a definite range of lipophilicity (see also p. 114) is essential for the insecticidal action of pyrethroides [39], although there is no correlation of lipophilicity with toxicity [40]. However, penetration through the nerve tissue to the target site within the nerve correlates with lipophilicity [41]. The more polar compounds like pyrethrin II or tetramethrin act faster in terms of more or less transient knock down of the insect.

The first synthetic variations, the first pyrethroids, were made by Staudinger and Ruzicka in Zürich more than 70 years ago, when they elucidated the chemical nature of pyrethrins as esters and synthesized chrysanthemic acid for the first time [42], independent of the contemporary works of Fujitani and Yamamoto [42a] in Japan. They could show, that certain substituted benzylalcohols and longer chained unsaturated aliphatic alcohols yield moderately active chrysanthemats [43]. Their results were commercialized shortly after by other people in the USA. Some of Harvills longchained saturated $C_{12} - C_{14}$ aliphatic chrysanthemats proved to be active against some aphids [44].

The first very active synthetic pyrethroid however, was very close in structure to the natural lead compound pyrethrin I. This compound allethrin was first synthesized by Schechter and LaForge [45, 46], and after almost 40 years in use, is still an interesting household insecticide.

In 1949 Synerholm started again, where Staudinger had left off, looking for active substituted benzylesters]47], while Barthel at the Boyce-Thomson-Institute found a quite interesting benzylester barthrin [48]. When M. Elliott entered this field of research in the 1950's the search for substituted aryl- and heteroaryl-methylchrysanthemates broadened considerably. At the same time the first variation of chrysanthemic acid by Farkas, Šorm and Kuřim [49] yielded, in 1957, permethric acid, whose importance was nor recognized at that time, were the dimethylvinyl side chain was exchanged for a dichlorovinyl moiety.

A first breakthrough in terms of higher activity was the discovery of the N-hydroxymethyl tetrahydrophthalimid and of 3-hydroxymethyl-5-benzylfuran as an alcohol component by Kato (1964) [50] and Elliott (1967) [51] respectively. However, the pyrethroids gained great commercial importance as agricultural insecticides only after the contemporary discovery of m-phenoxybenzyl alcohol by Itaya and Elliott (1968) [54], when Elliott combined this alcohol with another photostable component, the above mentioned Farkas acid, yielding permethrin and cypermethrin. An α-cyano-group improved the activity further, as discovered by Matsuo 1971 [55], so did a 4-fluoroatom (Fuchs 1976) [58] and the substitution of one chlorine atom by a CF_3 group (Huff 1977) [58b]. Particulary Elliott's deltamethrin (1974) [52], a single optically active stereoisomer, proved to have an hitherto unheard of insecticidal efficacy. At the same time the structurally quite different, but isosteric α-isopropylphenylacetic ester, fenvalerate, was discovered by Ohno (1972) [53] as a somewhat less active, but more easy to synthesize pyrethroid, having a novel structure, but the same action. This development was extended to the N-phenylvalinesters by Hendrik (1978) [60]. The insertion of a phenyl moiety into the $C-Cl$ bond of permethrin in connection with the introduction of a fluorine of the m-phenoxymethyl alcohol shifted the spectrum of activity for the first time towards ticks (Fuchs 1976) [59]. Another novel structure, having no close resemblance to the natural pyrethrin any more, was a hybrid, combining features of DDT-analogues and cypermethrin, synthesized by Holan (1977) [56]. The pentafluorbenzyl esters of Naumann (1976) [57] are of simpler structure, having much faster action against insects than any other pyrethroid. Similar parasubstitutet tetrafluorobenzyl esters prooved to be the first pyrethroids with activity in soil (McDonald 1979). Further variants were the biphenylmethanols (Plummer 1980) [63a, 159] and the phenylindanols as alcohol components (Engel 1980) [63, 160]. The ensemble of structural diversity was extended further by the exchange of the ester partial structure by an oxime (Bull 1978) [61]. More recent variations of the central part by Udagawa et al. [60] led to the ether type of pyrethroids (1980), or ones with hydrocarbon linkages (1982) which show an astonishingly high activity, which was not expected on the ground of structural experience in the 1970's. The latest and farthest deviation from the original pyrethrin I still showing insecticidal activity are ethers without the phenyl moiety, which is important for activity in the fenvalerate series, leading to the 3-phenoxy-4-fluorine-benzyl-neopenthyl ether.

Structural variations over 30 years culminated in insecticidal structures, which bear, at the first glance almost no resemblance to the active natural archetypical compound. However, as will be shown later, the very close resemblance to pyrethrin I is still retained in the shape and geometrical distribution of physical properties on the surface of the different pyrethroid molecules.

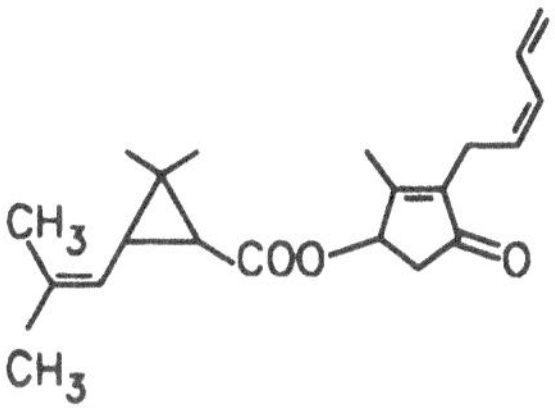

1R, trans, S
Nature

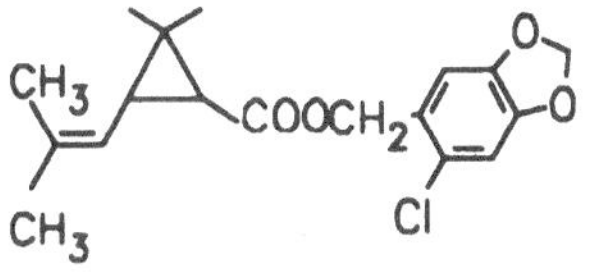

1957

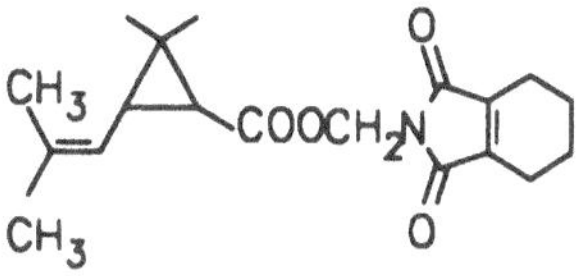

1962

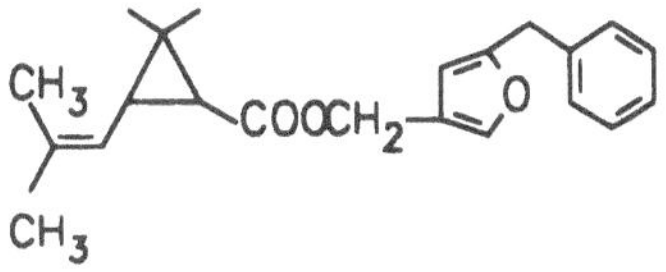

1965

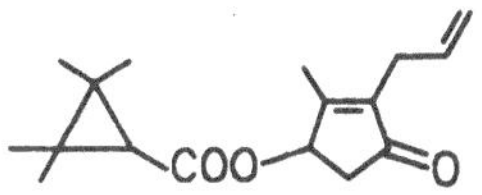

1966

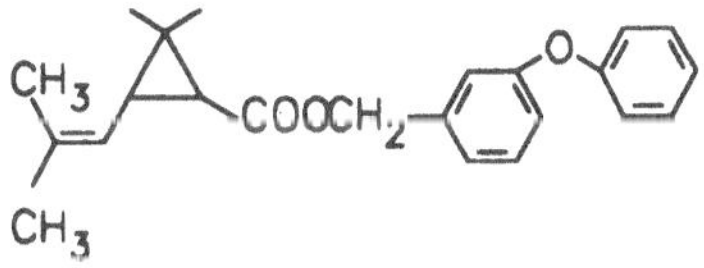

1968

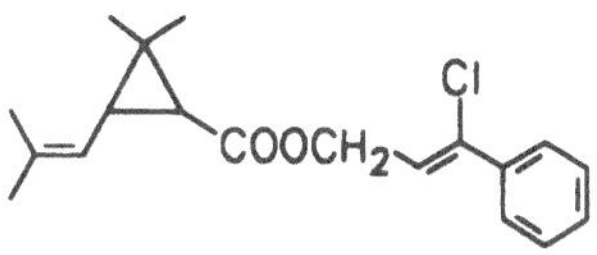

1969

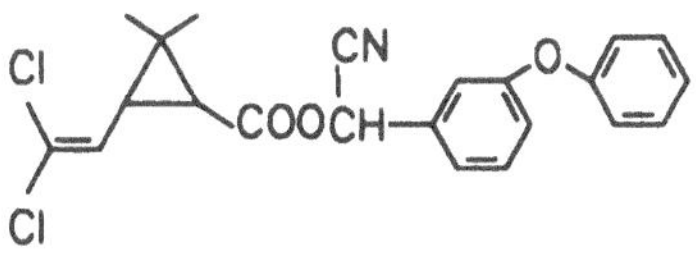

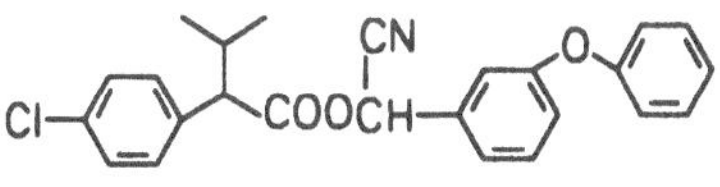

1971

1972

1971   1974

1973   1976

1976   1977

1977   1980

1980

1980   1980

1981   1982

1983   1983

1984

As far as they have been investigated, the active compunds, such as the oxims and the ether- and DDT-pyrethroids, deviating a great deal from the natural archetypical compound, act like phenothrin [64, 65, 66, 67] from the electrophysiological point of view.

This migration through this diverse field of active structures (an even more diverse picture is given by the collection of patent publications Vol. II) however is more like a walk through a swampy area on a narrow path, since each of the very active compunds is surrounded by a number of less active and many more inactive ones. Often a minute variation in structure, substituents and stereochemical shape of the starting compund leads to an inactive molecule. Therefore a look at inactive compounds is also interesting in order to understand the essential idealized molecule and its active conformation, which is supposed to be the cause of the typical pyrethroid actions (if there is anything like 'the typical pyrethroid', and we will not deal with different classes on the molecularpharmacological level, which happen to have common partial structures).

Next is a list of close structural variations of highly active pyrethroids having no or very low insecticidal activity:

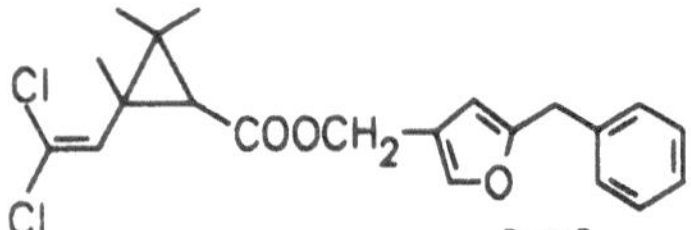

[73]

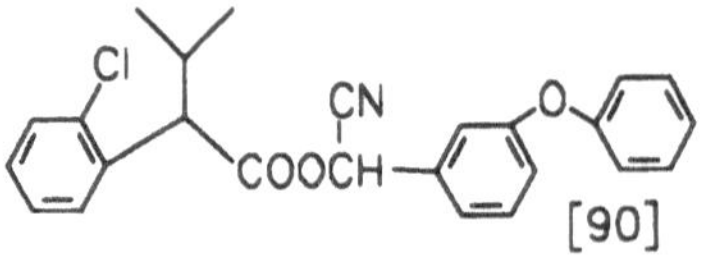

[90]

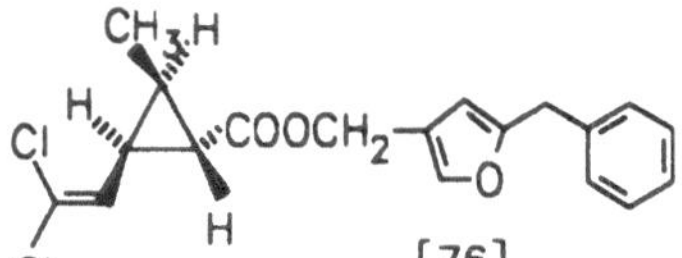

[76]

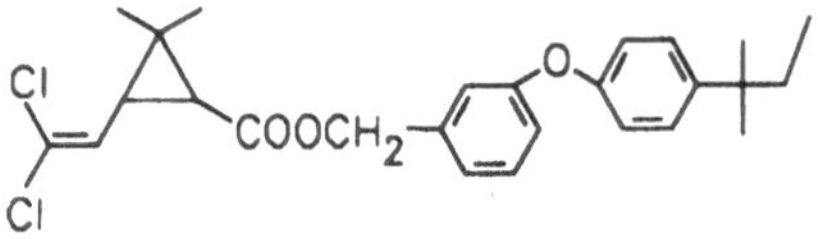

[72]

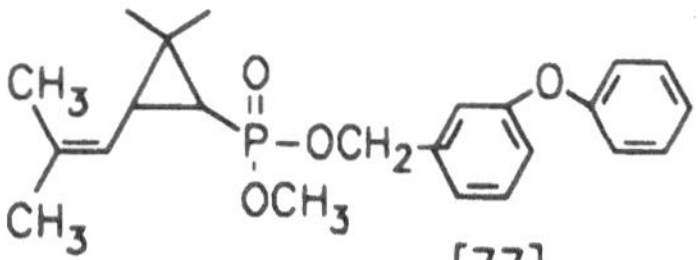

[77]

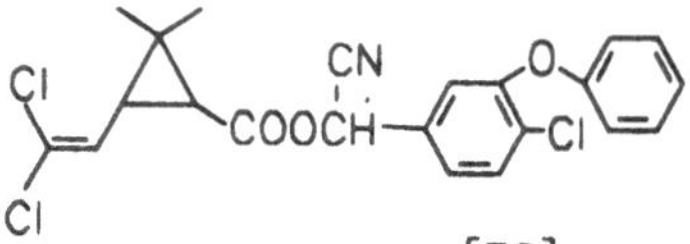

[78]

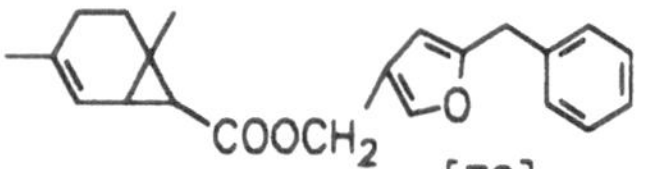

[79]

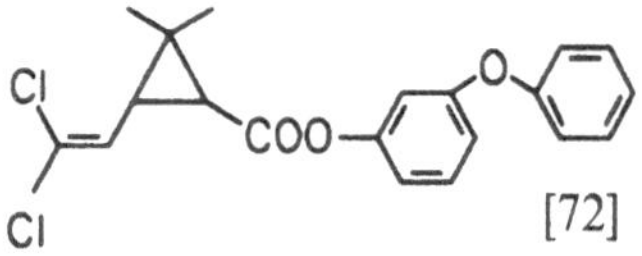

[72]

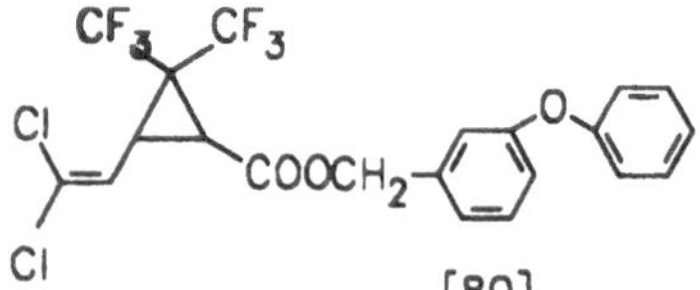

[80]

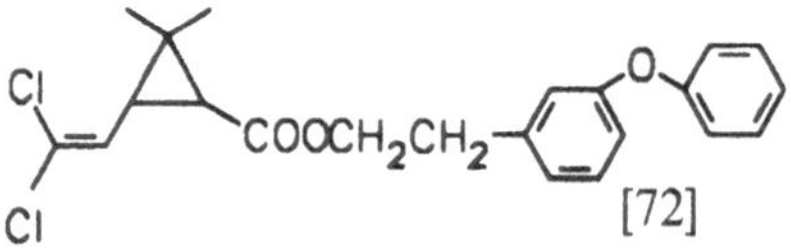

[72]

Structurally remote variations of highly active synthetic pyrethroids retaining moderate or higher activity against certain or many insects:

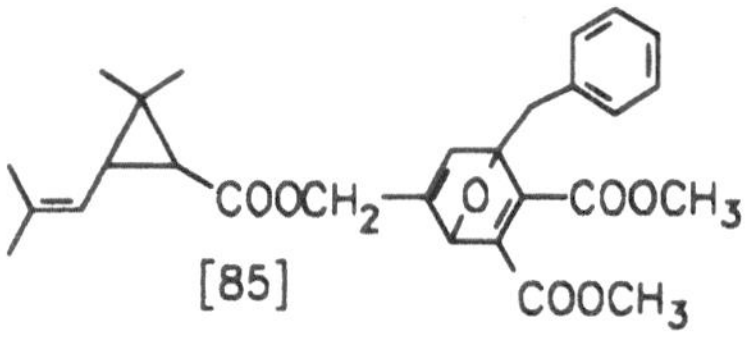

[85]

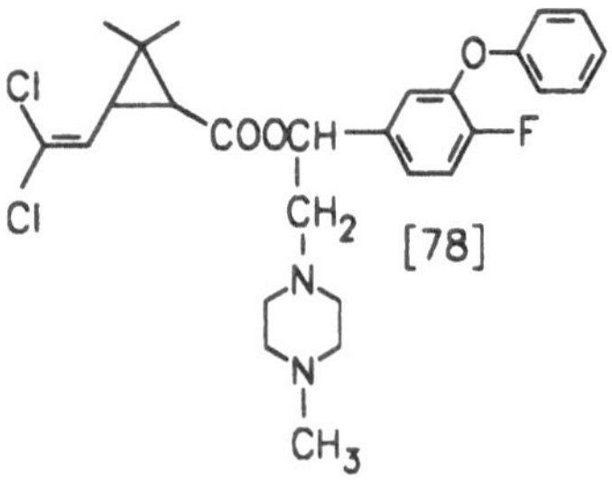

[78]

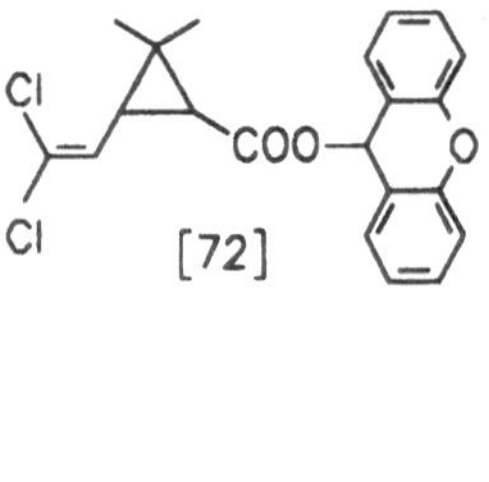

[72]

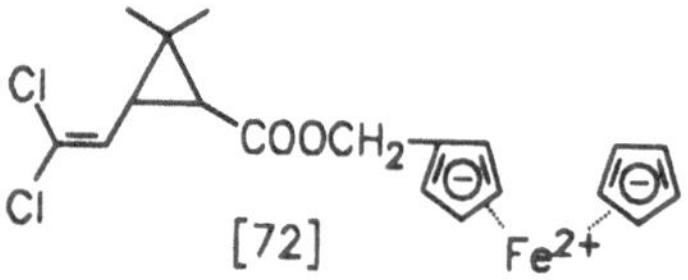

[72]

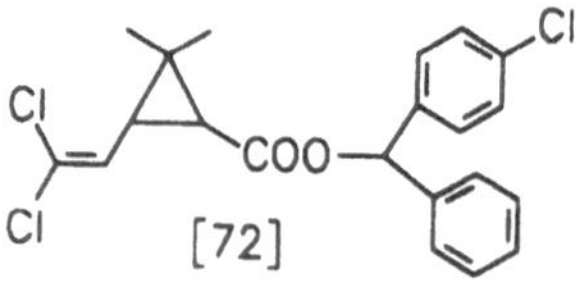

[72]

A common description of the active compounds would touch on the question of identity of pyrethroids. Structurally in the broadest terms, the field, were active pyrethroids can be found, may be described by the general formula

$$\pi - A - D - \pi'XS$$

π, π' = π-system (aromatic, olefinic), optionally substituted by further lipophilic substituents

A   = a small lipophilic anchor group of a limited size of a geminal dimethylmethylene as part of a cyclopropane system, or of an ether chain; or a isopropyl group attached to a neighbouring benzylgroup. Steric properties in this region are dominant. Electronic factors are less important.

D   = is a spacer group comprising the length of three atoms as an ester, ether, oxim or hydrocarbon alignment.

S   = Substituent containing a further π-system, preferentially in the metaposition of the benzylic π'-system

X   = may be there or not, a flexible pivot atom like oxygen, methylene and the like.

In the most cases, 'very active' acids give the best esters with any 'active' alcohol and vice versa. However, there are some examples, where certain alcohols, like pentafluorobenzyl alcohol exhibit only interesting qualities in the case of cyclopropanecarboxylates [72] and oxims of special aryl-haloalkyl ketones [66].

Almost all active alcohol components are of benzylic or allylic nature, having at least one C−H-bond left in the α-carbon, bearing the oxygen of the alcohol. The one exception is the 2-indanylpyrethroids. Phenol esters are altogether inactive.

An economically important, until now not well understood observation is the α-cyano effect, which only in the case of 3-phenoxybenzylalcohol and in the somewhat less active 3-allyl-substituted benzyl alcohol [87] (particularly 3-(Z)-3-chloroallyl) [88], improves the insecticidal activity considerably, but only if this cyanohydrin has the S-configuration.

The 4-phenoxybenzyl-isomer and most other meta-substituted-benzylalcohols lose their activity more or less by the introduction of the CN-group [89, 90]. Different effects are observed for the α-ethynylgroup in benzyl alcohol. In the meta-allyl-substituted-benzyl esters the R-isomer is twice as active as the S-isomer, as expected. However, in the case of the active isomer of the para-allyl-substituted-α-ethynylbenzyl esters, which has the same activity as the meta-isomer, the (S)-configuration is needed for activity [91], being much more active than the R-isomer.

In the course of time, several substituent effects in the very active 3-phenoxy benzyl-compounds were discovered, which raised the activity even more [92] as shown in Table 2.

Table 2. Changes in arthropodicidal activity by substituent exchange in phenothrin (±) cis/trans

Substituents				Median effective lethal dose against insects and ticks		Improvement factors	
X′	X″	Y	Z	Insects	Ticks	Insects	Ticks
CH$_3$	CH$_3$	H	H	50–100	>1000	+2–10 fold	+1–2 fold
Cl	Cl	H	H	10	>1000	+3–8 fold	+1–2 fold
Cl	Cl	CN	H	2–10	500–1000	+>2 fold	+10 fold
Cl	Cl	CN	F	1–5	500	−10 fold	+10 fold
Cl	4-ClC$_6$H$_4$	CN	F	50	5	−10 fold	−10 fold
Cl	4-ClC$_6$H$_4$	CN	H	100–500	100		−30 fold

The increase in activity by the 4-fluorine-substitution varies considerably both with acid component and insect species, for example for the geminal fluorochloro analogue of cycloprothrin [93]. Against *Heliothis punctigera* this bonus has about a factor of ten, while there is almost no significant increase against *Blatella germanica* or *Lucilia cuprina*.

Currently the ranking of pyrethroids having the highest intrinsic activity is as follows.

a) Against agricultural pests:

1R-cis, Z, αS

1R-cis αS

1R-cis, αS

[94]

1R-cis, αS

b) Against houseflies and mosquitoes:

1R-trans

1R-trans αS

1R-trans αS

c) Against spider mites:

1R-cis Z

d) Against cattle-ticks:

1R-trans, Z, αS

e) Against soil-insects:

1R-cis Z

A recent result of pyrethroid research, which is of particular interest, are the non-ester pyrethroids. They have a certain structural similarity to fenvalerate but with significant changes in the optimal position of the geminal dimethyl group. The 3-phenoxybenzyl-moiety is necessary for activity. Substituents in the ortho-position of the other aromatic ring eliminates activity. In contrast to fenvalerate an ethoxy group in the 4-position seems to be the optimal substituent. The influence of fluorine in the 4-position of the diphenylether system on activity differs very much within different species.

Generally, the influence of a fluorine substituent introduced in any position of a molecule on biological activity is of particular interest. While only minor steric changes are observed, a fluorine causes a very strong change in the electronic framework in the vicinity and sometimes more distant parts of the molecule, as well as distinct modification of physicochemical properties of the molecule.

Some structure-activity relations for nonester pyrethroids are shown next [95]:

active

inactive

As to the biological properties, the lack of fish-toxicity of this ether-type of pyrethroids is most interesting for their applicability in paddy fields and other irrigated crops.

Not only the carbonyl function is exchangable by a methylene group. The ester function altogether can be substituted by a saturated or unsaturated ethylene linkage to give the even more simplified hydrocarbon-pyrethroid or its structurally more demanding olefinic analogue. While the former is one of the few pyrethroids with no isomerism, only the trans-isomer of the latter one is insecticidally active (Table 53, p. 57).

2.2 Collection of Structure-Activity-Data

For a detailed analysis of structural dependencies of insecticidal and neurochemical activity, biological data of a variety of pyrethroids are necessary. The scientific literature of the last 15 years provides a wealth of data in a singular way. The most typical ones were selected and, if necessary, recalculated and expressed in terms of *relative toxicity* in order to make them comparable. Absolute insecticidal results of tests, repeated in other laboratories, or even in the same one, may differ by up to one order of magnitude, and are hardly ever exactly reproducible, even if extreme care is taken to minimize any of the natural and methodical fluctuations. However, the values relative to a standard control compound tested in the very same test at the same time are much more reliable and reproducible.

These relative insect toxicities are very usefull for further optimization of a series. Relative values obtained by careful topical application in a standardized procedure are suitable for quantitative structure-activity-correlations.

In the following tables of relative insecticidal data a high figure represents high insecticidal activity or toxicity. This corresponds to a low lethal dosis of active ingredient or absolute toxicity as LC_{50-100} (weight of active ingredients per spatial dimension: %, ppm, g/ha, mg/m^2, µg/l) or LD_{50} (weight of active ingredients per mass-unit of animal: mg/kg, µg/g, ng/insect).

For example if the relative toxicity of a compound A = 100, of B = 10 and of C = 500, it means that C is 50 times more toxic than B and five times more toxic than A. Accordingly the relative speed of action is expressed; a higher figure means faster action.

The following structure-activity-data are grouped together according to the type of structural variations in question.

2.2.1 Structure-Activity-Data of Ester-Pyrethroids

2.2.1.1 Variation of the Acid Moiety

Table 3. Relative toxicities of esters of Elliott-alcohol (Bioresmethrin = 100)

R—OCH₂ ...

Acid		Relat. Toxicity		Lit.
		Musca d.	*Phaedon c.*	
	1Rtrans	150	160	[96]
	1Rtrans, E	39	52	[98]
		74	30	[98]
	1Rtrans	160	160	[96]
	1Rtrans	300	390	
	1Rtrans	130	130	
	1Rtrans	86–250	240–250	[98, 99, 100]
,,	1Rcis	200–280	220–290	[96, 99]
,,	1Strans	5		[98]
,,	1Scis	14	13	[99]
	1Rtrans	390	180	[98, 99]
	1Rcis	260	200	[100]
	1Rtrans	210	270	
,,	1Rcis	360	250	
Cl—C≡C	1Rtrans	32	18	[100]

Table 3. (continued)

Acid		Relat. Toxicity		Lit.
		Musca d.	*Phaedon c.*	
	(±) cis/trans	2	1	[101]
	(±) cis/trans	8	1	
	,,	10	1	
	,,	23	50	
	,,	70	60	
	,,	—	—	
	,,	100	190	
	,,	54	85	

Table 4. Relative toxicities of Elliott-alcohol esters

Acid		Rel. Toxicity				Lit.
		Musca	*Phaedon*	*Anopheles*	*Aëdes*	
	(±(cis/trans Resmethrin	≡ 100	100	100	100	[51]
	1Rtrans	31	120			

Table 4. (continued)

Acid		Rel. Toxicity				Lit.
		Musca	*Phaedon*	*Anopheles*	*Aëdes*	
	1Rtrans	330	460	250	310	
		190	90	130	110	
		7	52			
	(−)	5	22	51	64	
	(+)	4	12	30	29	
Pyrethrum		≡ 100				[102]
	(−) 1Rtrans	420				[103
„	(+) 1Strans	48				
„	(+) 1Rcis	310				
„	(+) 1Scis	64				
„	1Rtrans	30				
„	1Rcis	120				

Table 5. Toxicity of stereoisomers of resmethrin and permethrin against fly and mustard beetle

	Stereochemistry	Housefly		Mustard beetle	Lit.
		ng/fly	rel. Tox.	rel. Tox.	
Resmethrin	(±) cis/trans	50	100	100	[19]
	1Rcis	40			
(X = CH₃	1Rcis	4000			
Y = CH₂	1Rtrans	13			
Z = O)	1Strans	1770			

Table 5. (continued)

	Stereochemistry	Housefly		Mustard beetle rel. Tox.	Lit.
		ng/fly	rel. Tox.		
Permethrin	$(\pm)$ cis/trans	60		100	
(Y = O	1Scis		160–320	11	[98]
Z = −CH=CH−	1Rcis		160–320	140	[99]
X = Cl)	1Strans		1	12	[100]
	1Rtrans		74–88	210–630	[104]
					[105]
					[106]
X = F	1Rcis		170	120	
	1Rtrans		94	42	
X = Br	1Rcis		180	360	
	1Rtrans		78	290	

◆**Table 6.** Insect toxicity of isomeric pyrethric esters [107]

X	R	Stereo chemistry	Rel. Toxicity					
			Musca d.			Blattella g.		
			Y = CH		Y = N	Y = CH		Y = N
			Z = H	Z = F	Z = H	Z = H	Z = F	Z = H
CH$_3$	CH$_3$	1Rtrans, E	7			<1		
		1Rtrans, Z	2			<1		
		1Rcis, E	2			<1		
		1Rcis, Z	4			<1		
H	CH$_3$	1Rtrans, E	2			<1		
		1Rtrans, Z	2			<1		
		1Rcis, E	2			<1		
		1Rcis, Z	29	84	6	27	45	1
	C$_2$H$_5$	1Rcis, Z	68	97	16	32	54	25
	CH$_2$CF$_3$	1Rcis, Z	120		29	29		33
	nC$_3$H$_7$	1Rcis, Z	89		17	11		30
	i-C$_3$H$_7$	1Rcis, Z	82	103	12	42	59	19
	◁	1Rcis, Z	93	139	98	27	81	54
	⟨CF$_3$⟩(CF$_3$)	1Rcis, Z	49	126	98	74	112	
	⊲	1Rcis, Z	47	102		52	21	
	Deltamethrin		100			100		

3*

Table 7. Influence of halogen on relative insecticidal activity of dihalovinylcyclopropanecarboxylates [104] (bioresmethrin $\equiv$ 100)

Hal	Stereo-chemistry	X = H		X = CN			
		Housefly	Mustard beetle	Housefly		Mustard Beetle	
				αS	αR	αS	αR
F	1Rcis	170	180	350	3	2000	110
	1Rtrans	94	42	170	8	330	21
Cl	1Rcis	210	320	2400	82	5500	170
	1Rtrans	88	630	1400	81	2200	110
Br	1Rcis	180	360	2800	48	5500	410
	1Rtrans	78	290	1140	39	2400	21

Table 8. Influence of sidechain substitution in cyanophenothrin on toxicity to *Musca d.* and *Phaedon c.*

X	Y	Stereochemistry	Rel. Tox.		Lit.
			Musca d.	*Phaed.*	
CH$_3$	CH$_3$	±c/tRS	100	100	[68]
Cl	Cl	,,	150–180	600	[100, 110]
CF$_3$	CF$_3$	,,	40		[111]
Cl	CF$_3$	±c/tZRS	330		
Br	CF$_3$	,,	310		
F	CF$_3$	,,	170		
H	CF$_3$	±c/tERS	8		
Cl	C$_2$F$_5$	±c/tZRS	150		
Cl	CF$_2$Cl	,,	80		
Cl	CF$_3$	1RcisZ, αS	3800		
CF$_3$	Cl	,, E ,,	>3800		
Br	Br	1RcisαS	1500	7600	
Cl	Cl	,,	1300	7600	

Table 9. Relative activity of flumethrin-isomers against flies (deltamethrin = 100) [112]

	X	Toxicity
1RcisE	H	14
	F	22
1RcisZ	H	<1
	F	1
1RtransE	H	<1
	F	1
1RtransZ	H	37
	F	59

Table 10. Toxicities of vinylogous esters of cypermethrin [113]

X	Y	LC$_{50}$ (ppm)		
		Musca	*Leptinotarsa*	*Spodopt.*
Cl	Cl	10	10	10
Cl	CH=C(H)(Cl)	10	10	10
Cl	CH=C(Cl)(Cl)	200	70	70
H	CH=(Cl)	10	10	10

Table 11. Absolute toxicities of highly active pyrethroids (LD$_{50}$)

	Anopheles stephensi ng/insect	Glossina aust. ng/insect	Boophilus Adults ♀ µg/g	Larvae µg/l	Musca dom. mg/kg	Blattella mg/kg
Deltamethrin	0.036	0.08	0.3	29	0.03	0.05
Cypermethrin	0.2	0.75	2.0	120		
Fenpropanate	0.72	2	3.0	200		
Fenvalerate	1.5	8.5	5.2	380	1.6	0.74
Permethrin	1.8	2.3	2.1	230	1.0	0.5
Phenothrin	4.0	10				
Literature	[114]		[115]			[14]

Table 12. Absolute toxicity after topical application of 3-phenoxybenzyl-pyrethroids

$$R\!-\!O\overset{X}{\diagup}\!CO\text{—(3-phenoxybenzyl)}$$

R	X	Stereochem.	Toxicity LD$_{50}$ Musca ng/fly	Heliothis mg/kg larvae	Locust mg/kg insect	Lit.
(2,2-dimethyl-cyclopropane-CO)	H	–	74			[64]
(2,2-dichlorovinyl-dimethylcyclopropane-CO)	H	± cis/trans	800 / 31	0.018 µg/larva	130	[90]
(2-dimethyl-propenyl-dimethylcyclopropane-CO)	,,	,,	77	0.22 µg/larva		
	CN	,,	210		100	
	,,	1RcisS	5		25	
	,,	1RtransS	3			
	,,	1RtransR	72			
(Cl/CF$_3$-vinyl-dimethylcyclopropane-CO)	,,	(±) cis/trans	1.7	0.0034 µg/larva		
(4-chlorophenyl-isopropyl-CO)	H	±	91			[60]
	CN	,,	42	12	10	[64]
	,,	S, S	11			[64]
	,,	SR	400			[64]

Table 12. (continued)

Structure					
[4-Cl-phenyl-NH-CO isopropyl]	H	±	210	55	
[4-Cl-phenyl-N(CH₃)-CO isopropyl]	"	"	360	500	
[tetrafluoro-isoindoline CO, isopropyl]	" CN	"	49 63	140 28	[60]
[isoindoline-CO isopropyl]	H	"	180	150	
[Br₂C=CH-cyclopropane-CO]	H "	1R, (opt.pure) 1S, (opt.pure)	*Periplaneta americana* 0.085 µg/g > 20 µg/g		[116]

Table 13. Insecticidal activity of benzospirocyclic pyrethroids [117, 118]

Compound	Stereo-chemistry	Activity (LD$_{50}$)	
		Housefly µg/♀	Tobacco cutworm ppm
[indane-spiro-cyclopropane-COO-CH(CN)-phenoxyphenyl]	±	0.0045	2
	1R; S	0.0013	0.6
[CH(CN)(CH₃)-phenyl-O-phenyl, F]	±	0.0044	1.8
[F-indane-spiro-cyclopropane-COO-CH(CN)-phenoxyphenyl]	±	0.0052	1.0

Table 13. (continued)

Compound	Stereo-chemistry	Activity (LD$_{50}$)	
		Housefly µg/♀	Tobacco cutworm ppm
Br—[structure] COO ,,	±	0.018	7.2
[structure] COO ,,	± cis	0.064	
	± trans	0.0061	1.8
[structure] COO ,,	±		5
[structure] COO ,,	± cis/trans	0.16	46
[structure] COO ,,	±	0.079	
Cl—[structure] COO ,,	±	0.0075	

Table 14. Relative toxicities of phenylisovalerate-pyrethroids [101, 119, 120] to

	Musca domestica	Aëdes aegypti L.	Spo-doptera littoralis L.	LD$_{50}$ mg/kg mouse
(±) RS, R′S′ Fenvalerate	= 100	100	100	245
S, S′	350–440	270	430	50
	LD$_{50}$ 11 ng/fly			
S, R′	2–5	29		> 600
S, R′S′	200	190		81
R, R′S′	—	—		> 5000

Table 14. (continued)

(±) RS, R'S'		89	
S, S'		440	290
S, S'		340	70

Table 15. Relative toxicities of phenylisovalericesters to housefly [102]

Compound	Rel. Toxicity	Compound	Rel. Toxicity
Pyrethrum	≡ 100		
	710		
	200		⁺ 2000
	50		480
			640
	370		290

Table 16. Absolute insect-toxicity LC$_{50}$ of some benzocondensed homologues of fenvalerate [121, 122] (leaf treatment; ppm)

Compound		Aphid	Mite	Southern armyworm	Mexican bean beetle	Housefly
	[121]	2	82	3	1	9

Table 16. (continued)

Compound		Aphid	Mite	Southern armyworm	Mexican bean beetle	Housefly
(structure)	"	0.3	inact.	10	3	90
(structure)	"	0.2	inact.	42	13	inact.
(structure)	"	1.5	inact.	165	5	inact.
(structure)	"	2	inact.	65	9	400
(structure)	"	5	inact.	250	23	inact.
(structure)	"	0.8	500	17	3	
(structure)	"	0.7	210	45	11	
(structure)	"	0.2	400	7	1	
(structure)	"	5	500	> 500	500	

Compound					µg/insect topical LD_{50}	
(structure)	"	[122]	19	12	0.09	19
(structure)	"		100	100	9	

Table 16. (continued

Compound		Aphid	Mite	Southern armyworm	Mexican bean beetle	Housefly
	,,	45	54	0.35		0.25
	,,	>100	>100	8		0.76
	,,			0.07		0.4
	,,			0.7		0.4

Table 17. LD$_{50}$-values of some N-phenylvalineester-pyrethroids [90]

	Topical LD$_{50}$		
	µg/larva *Heliothis vir*	µg/adult *Musca d.*	*Spodoptera*
	0.055	0.23	0.8
,, ,,	0.08	0.16	0.19
,,	0.018	0.075	
,,	0.066	0.45	0.18
,,	11	0.38	

Table 17. (continued)

| | Topical LD$_{50}$ | | |
	µg/larva *Heliothis vir.*	µg/adult *Musca d.*	*Spodoptera*
	0.85	0.33	0.08
		Only weakly active [72]	
	38	1	
	0.5	0.3	
"	0.15	0.13	
"	0.19	0.11	
"	5	5.5	
"	0.15	0.18	
"	3	0.6	
"	1	0.4	
"	0.6	0.6	

Table 18. Substituent effects in phenylisovalerates and homologous pyrethroids [122]

				LC$_{50}$ Topical µg/insect		
				Heliothis	Musca	Tetranychus
X	Y	Z	W			
H	H	H	H	0.7	0.34	> 1000
Cl	H	H	H	3.5	> 1.0	> 1000
F	H	H	H	0.5	0.4	29
F	H	H	CN	0.04	0.13	7
H	CH$_3$	H	H	23		
H	H	CH$_3$	H	4	5	

X	W		Heliothis	Musca	
H	H		1	0.2	
H	CN		0.09	0.2	
4-Cl	H		0.09	0.09	
4-Cl	CN		0.01	0.04	
2-Cl	CN		Inactive [151 a]		

Table 19. Activity of some seco-cypermethrin-homologues [123]

Compound		LC$_{50}$ (ppm) (synergized)				
		Aphid	Mite	Southern armyworm	Mexican bean beetle	Housefly
	± c/t	0.2	74	3	0.4	5
	,,	8 (0.8)	330	80 (24)	24 (18)	60
	,,	100	370	> 400	70	> 500
	,,	0.5 (0.2)	200	100 (40)	3 (0.2)	40

Table 19. (continued)

Compound	LC$_{50}$ (ppm) (synergized)				
	Aphid	Mite	Southern armyworm	Mexican bean beetle	Housefly
[structure: Cl₂C=CH–CH(iPr)–COO] "			inactive		
[structure: Cl₂C=CH–CH(CH=CH₂...)–COO] "	4	250	> 500	20	500
[structure: Cl₂C=CH–CH(CH₂CH=CH₂)–COO] "	4	120	60	9	70
[structure: Cl₂C=CH–CH(cyclopropyl)–COO] "	0.7 (0.1)	250	30 (3)	3 (0.7)	40
[structure: 4-Cl-C₆H₄–CH(iPr)–COO] "	0.5 (0.05)	140	5 (1.2)	0.7 (0.09)	8

Table 20. Relative toxicities of 1-phenylcyclopropancarboxylate-pyrethroids to housefly

Pyrethroid	Rel. Tox. Musca [124]
Permethrin LD$_{50}$ 5 ng/fly	$\equiv 100$
[structure: ethoxyphenyl-cyclopropane-COO-CH₂-(3-phenoxyphenyl)]	
" ($\pm$)	3
" ($+$)	5
" ($-$)	3

Table 20. (continued)

Pyrethroid		Rel. Tox. Musca [124]

"" (±) 26

LD$_{50}$ 75 ng/fly
LD$_{50}$ 20 ng/blowfly

"" "" (+) 3

"" (−) 66

"" "" (+) 14

"" (±) 23

(±) inactive [69]

"" (±) inactive [77]

Table 21. Toxicities of 1-phenyltetrafluorocyclobutanecarboxylate-pyrethroids [125]

X	Y	Abs. config. of		LD$_{50}$ (mg/kg)		
		Acid	Alcohol	Blowfly		Cockroach
				topical	synergized	topical
EtO	CN	R	R	4	0.15	24
"	"	S	S	0.36	0.006	1.6

Table 21. (continued)

X	Y	Abs. config. of		LD_{50} (mg/kg)			
		Acid	Alcohol	Blowfly			Cockroach
				topical	synergized		topical
,,	,,	S	R	5.2	0.08		28
,,	,,	R	S	135	1.7		>1000
,,	C≡CH	S	R	0.18	0.005		1.8
,,	,,	R	R	28	1.9		195
,,	,,	S	S	108	22		560
Cl	,,	S	R	0.12	0.013		1.4
Deltamethrin				0.24	0.002		0.45

Table 22. Effect of fluorine substituents on biological efficacy of pyrethroids [126]

Compound		Substituent-variation	Biol. Effect	Comparison of efficacy	Lit.
	(±)	X = Y Cl, F	Insect.tox.	Cl > F	
	1Rcis	Cl, F	Mammal.tox.	Cl ≪ F	
	±	X = Cl Y = CF₃	Insect.tox.	Cl < CF₃	
		X=H, F; Y = H	Insect.tox.	H < F	
		X = H, Y = F, H	Insect.tox.	H > F	
		X = CH₃, CF₃	Insect.tox.	CH₃ ≫ CF₃	[247]
	,,	X = Cl, CF₃	Insect.tox.	Cl ~ CF₃	[126a]
	,,	X = CH₃, CF₃	Insect.tox.	CH₃ < CF₃	
	,,	X = Cl, CF₃	Insect.tox.	Cl < CF₃	[127]
	,,	X = Cl, F	Insect.tox.	Cl < F	

Table 22. (continued)

Compound	Substituent-variation	Biol. Effect	Comparison of efficacy	Lit.
	X = CH$_3$, CF$_3$ X = H, F	Insect.tox. Insect.tox.	CH$_3 \gg$ CF$_3$ H $\gg$ F	
,,	X = H, F	Insect.tox.	H < F	
,,	X = H, F	Insect.tox.	H < F	[151]
	X = H, F	Insect.tox.	H ~ F	
	X = H, F	Insect.tox.	H $\ll$ F	[128]

Table 23. Relative toxicity of some pyrethroids in comparison to parathion

	Musca domestica	Phaedon cochleariae	Plutella maculipennis	Spodoptera littoralis	Heliothis zea	Tetranychus urticae	Lit.
Parathion	100	100	100	100	100	1,00	[129]
	300						
,,	12						
,,	60						
,,	2						

Table 23. (continued)

	Musca domestica	Phaedon cochleariae	Plutella maculipennis	Spodoptera littoralis	Heliothis zea	Tetranychus urticae	Lit.
(structure)	26						
(structure)	60						
" (structure)	54	10	5	570		30	[130]
" (CN structure) "	100	60	10	1500		24	
" (structure) "	80	30		650		10	
(structure)	30	12		2000	620		
" (CN structure) "	850	110		2000	1320	5	
(structure)	35			110	150	70	[131]
(structure) "	20			100	—	24	
(structure) "	20			60	380	10	
(structure) "	40			210	280	15	

Table 23. (continued)

	Musca domestica	Phaedon cochleariae	Plutella maculipennis	Spodoptera littoralis	Heliothis zea	Tetranychus urticae	Lit.
`,,`	13			20	—	—	
	370			110	310	30	

Table 24. Absolute insecticidal activity of some household-pyrethroids [132]

	% Active isomers	Topical LD_{50} µg/insect			
		Musca d.	Culex pip.	Aedes	Blattella
d-Prallethrin	100	0.054	0.003		0.23
d-Tetramethrin	100	0.22	0.016		2.6
d-Allethrin	50	0.24	0.037		2.7
Allethrin	25	0.38	0.067	0.012	12.0
Pyrethrines	100	0.85	0.022	0.003	0.6
Terallethrin	50	0.53	0.064	0.018	5.7
d-Phenothrin	100	0.022	0.0075		0.9
Cyphenothrin	50	0.026	0.0028		0.16
Permethrin	50	0.027	0.0072		0.39
Cypermethrin	25	0.012			

2.2.1.2 Variation of Alcohol Moiety

2.2.1.2.1 Structurally Flexible Alcohols

Table 25. Relative toxicities of some chrysanthemates against housefly and mustard beetle

R	Stereochemistry	Rel. Tox.		Lit.
		Musca	Phaedon	
	1Rtrans bioresmethrin 1Rcis	100 (LC_{50} 10^{-4} M) 41	100 30–64	[99]

Table 25. (continued)

R	Stereochemistry	Rel. Tox.		Lit.
		Musca	*Phaedon*	
	1Rtrans, RS	1	1	[133]
	1Rtrans, RS	16	13	
	1Rtrans	30	70	
	1Rtrans, RS	150	96	
	1Rtrans, S	2	160	[96]
	1Rtrans, S	10	4	
	1Rtrans, S	10	30	
	(±) cis/trans	15	52	[108]
	(±) cis/trans	7	1	

Table 26. Relative toxicities of some heterocyclic chrysanthemates

	Musca dom.	*Phaedon cochl.*	*Anopheles*	*Aëdes*	Lit.
Standard: Resmethrin					
(±) c/t	100	100	100	100	

Table 26. (continued)

		Musca dom.	Phaedon cochl.	Anopheles	Aëdes	Lit.
Resmethrin-isomers						
	1Rtrans	240	270	150	190	[51]
	1Rcis	99	140	220	300	
	(±) c/t	12	5			
,,		42	10			[134]
,,		24	23			
,,		46	43			
,,		13	15			
,,		1				
,,		46				[51]
,,		—				[134]
,,		27	16			
,,		7	5			[135]

Table 27. Relative toxicities of some chrysanthemates to housefly [136–138]

	Rel. Tox.		Rel. Tox.
Standard Pyrethrum	100	Standard pyrethrum	
	400	,,	100

Table 27. (continued)

	Rel. Tox.		Rel. Tox.
	200		10
	700		10
	50		300
	inact.		1500
	80		1400
	240		100
	560		300
	10		64
	40		440
	90		1280
	100		

Table 28. Effect of α-substituents on rel. insect. toxicity of 3-phenoxybenzyl-pyrethroidesters [139] in

		Housefly (α-effect)		Mustard beetle (α-effect)	
Y	Z	X=CH$_3$ 1Rtrans	X=Br 1Rcis	X=CH$_3$ 1Rtrans	X=Br 1Rcis
H	O	≡100	≡100	≡100	≡100

Table 28. (continued)

		Housefly (α-effect)		Mustard beetle (α-effect)	
CH$_3$	O	13	23	21	22
C≡CH	O	40	45	140	288
CN	O	160	450	220	1120
		(1.6)	(4.5)	(2.2)	(11.2)
H	CH$_2$	≡100	≡100	≡100	≡100
CN	CH$_2$	150	840	190	80
		(1.5)	(8.4)	(1.9)	(0.8)
H	CO	≡100	≡100	≡100	≡100
CN	CO	310	63	470	540
		(3.1)	(0.6)	(4.7)	(5.4)
H	C=CH$_2$	≡100	≡100	≡100	≡100
CN	C=CH$_2$	21	14	97	110
		(0.5)	(0.14)	(1)	(1.1)
H	–	≡100	≡100	≡100	≡100
CN	–	13	23	3	11
		(0.13)	(0.23)	(0.03)	(0.11)

Table 29. Effect of α-substituents on rel. insect. toxicity of 3-phenoxybenzyl-pyrethroidesters [139] in

	Housefly (α-effect)		Mustard beetle (α-effect)	
Y	X = CH$_3$ 1Rtrans	X = Br 1Rcis	X = CH$_3$ 1Rtrans	X = Br 1Rcis
H	≡100	≡100	≡100	≡100
CH$_3$	0.2	4	<0.1	1
C≡CH	1.8	22	4	15
CN	15	9	12	17
	(0.5)	(0.09)	(0.12)	(0.17)

Table 30. α-Substituent-effect in 3-phenoxybenzylpyrethroids on relative toxicity to housefly and mustard beetle

		X = CH$_3$ 1Rtrans		X = Br 1Rcis		Lit.
Y	Z					
		Musca	*Phaedon*	*Musca*	*Phaedon*	[140]
H	H	≡100	≡100	≡100	≡100	
CN	H	170	180	350	333	

Table 30. (continued)

Y	Z	X = CH$_3$ 1Rtrans		X = Br 1Rcis		Lit.
CH$_3$	H	12	18	28	10	
CF$_3$	H	3	6	16	7	
F	H	6	30	17	0.3	
C=CH$_2$	H	33	40	44	36	
$-C\equiv CH$	H	44	120	61	139	
$-C\equiv C-CH_3$	H	14	36	67	50	
$-C\equiv C-CH=CH_2$	H	17	14	4	19	
3-Furyl	H	0.5	16	11	4	
3-Pyridyl	H	0.5	2	1	1	
C$\equiv$CH	CH$_3$	— inactive —				
$-CH_3$	CH$_3$	— inactive —				
CN	CH$_3$	— inactive —				[74]

Table 31. α-Substituent-effects on insecticidal activity of pyrethroids against housefly

Compound	Order of activity X	Lit.
	C$\equiv$CH > CN > H	[141]
	C$\equiv$CH > CN > H	
	CN > C$\equiv$CH > H	[130]
	inactive	[74]
	H $\gg$ CH$_3$ $\ggg$ CN	[72]
	H $\gg$ CN $\ggg$ CH$_3$	[139]

Table 32. αCN-effect $\left(\dfrac{CN}{H}\right)$ in two insect species of different alkylphenylacetates (x = CN or H)

R		Musca dom.		Phaedon coch.	
		Alkyl		Alkyl	
		iC_3H_7	cC_3H_5	iC_3H_7	cC_3H_5
4	Cl	3.2	1.5	4.5	2.9
4	Br	3.9	1.2	7.2	2.7
4	F		0.9		3.5
4	CH_3	2.8	1.5	18.3	2
3, 4	CH_3, CH_3		1.5		9
3, 4	OCH_2O		0.7		1.4
4	Cl				
		1.2	2.2	5	6.1
4	Cl				
			1		0.6

Based on relative topical toxicity as published in [142]

Table 33. Effect of α-substituents in phenoxybenzyl- and phenoxypyridyl-pyrethroid [90]

				Topical µg/insect.		
				Heliothis v. larva	Musca adult	Spodopt. larva
(±) cis/trans	X	Y	Z			
	O	H	CH	0.018	0.031	0.007
	O	CN	CH	0.0069	0.0065	0.001
(±) cis	O	CH_3	CH	0.025	0.029	
(±) trans	O	CH_3	CH	0.075	0.081	
(±) cis	O	H	N	0.025	0.052	
(±) trans	O	H	N	0.043	0.19	0.008
(±) c/t	O	CN	N	0.025	0.027	0.039
(±) cis	O	CH_3	N	0.016	0.071	
(±) trans	O	CH_3	N	0.035	0.22	
(±) c/t	S	CH_3	N	0.019	0.034	

Table 34. Influence of fluorine substituent on insecticidal activity (% mortality) [143]

Compound	Rate ppm	Insect.		
		Aphis fabae	*Majura vicea*	*Musca dom.*
	100 25 12.5	100 10 10	100 100 –	70 40 30
	100 25 12.5	– –	– –	40 30
	25 12.5	100 80	–	80
	6.3 3.1	100 90	100 90	90–100 –

Table 35. Comparison of toxicity against several insect species cypermethrin vs. vivithrin (= azabonus) [144] and cypermethrin vs. cyfluthrin (= fluorine-bonus) [78]

		LC$_{90}$ (ppm)					
		Tobacco budworm	Beet armyworm	Codling moth	Leaf-hopper	Whitefly	Res. housefly
Cypermethrin	X = CH Y = H	30	12	14	2o	80	14
Vivithrin	X = N Y H	32	12	3	5	60	7
Azabonus		1	1	3	4	1.3	2

$\dfrac{\text{LC}_{50}\text{ Cyfluthrin}}{\text{LC}_{50}\text{ cypermethrin}} =$	*Ceratitis*	*Phaedon*	*Plutella*	*Laphygma*	*Myzus*		*Euscelis*	*Tetrany*
Fluorinebonus	5	5	2	5	1		25	25

Table 36. Influence of position of fluorine substitution in cyfluthrin on toxicity to *Laphygma fr.* [145]

Position of F	LC_{95} (ppm)
2	50
4	1
5	5
6	100
none	5
2'	10
3'	25
4'	5

Table 37. Comparison of 3-phenylbenzyl pyrethroids and 3-phenoxybenzyl pyrethroids and analogues

		Rel. toxicity by topical (foliar) application (permethrin = 1)					
		Southern army-worm	Cabbage looper	Tobacco budworm	Mexican bean beetle	Milkweed bug	Lit.
	± c/t	1	1	1	1	1	[146]
,,	± c	2.4	1.1	1.8	2.0 (1.1)	2.7	
	,,	0.4	0.5	0.5	0.3 (0.1)	0.1	
	,,	0.4	0.3	0.5	0.3	0.8	
	,,	1.1	0.3	0.4	1 (0.6)	1.3	
,, ,,	± c/t	0.6					
,,	,,	inactive					

Table 37. (continued)

| | Rel. toxicity by topical (foliar) application (permethrin = 1) | | | | | |
	Southern army-worm	Cabbage looper	Tobacco budworm	Mexican bean beetle	Milkweed bug	Lit.
	1.87 (1.8)	2.2	1.3	2.4 (54)	11.6	
± c/t	0.8 (1.2)	2.1	1.3	1.6 (23)	4.2	
± t ,,	0.07					[147]
± c/t	0.9 (1)	1.5	1.2	2.9 (5.2)	13.9	
,,	1.3 (0.7)	3	4	0.7 (6.1)	4.1	
± c	0.01			0.05 (0.7)	<0.1	
,,	0.7			0.7	4.6	
,,	2.1			1.4	1.3	
,,	0.2			0.9	1.8	

Almost inactive to housefly and mustard beetle [148]

Table 38. Efficacy of perfluorinated benzyl esters against some insects [149]

Compound	Musca d. LD$_{100}$/min.	Lucilia c. LD$_{50}$ mg/kg synergized	Aëdes LC$_{100}$	Plutella LC$_{100}$ ppm
	0.2 ppm 120′		0.02 ppm 120′	10 ppm
	20 ppm 7^h		2 ppm 120′	50 ppm
	inactive		inactive	500 ppm
[66]		0.3		
		10		
		> 180		

Table 39. Activity of substituted tetrafluorobenzylester [150] against corn root worm larvae

R	Z	Stereoch.	LC$_{50}$ (ppm)[a]	LC$_{50}$ ppm in soil 5 weeks after treatment
4-CH$_3$	H	± cis-Z	0.02	0.3
4-CH$_3$	H	± trans	0.03	> 1
4-CH$_3$	H	1Rcis	0.007	∼ 0.25
4-CH$_3$	H	1Scis	1.1	> 10
4-CH$_3$	H	1Strans	0.05	> 1
4-CH$_3$	C≡N	± cis	inactive	
2-CH$_3$	H	± cis	0.08	
3-CH$_3$	H	± cis	0.1	
4-CF$_3$	H	± cis	> 5	
4-OCH$_3$	H	± cis	0.03	> 2

Table 39. (continued)

R	Z	Stereoch.	LC_{50} (ppm)[a]	LC_{50} ppm in soil 5 weeks after treatment
4-allyl	H	± cis	0.01	0.2
4-CH_2OCH_3	H	± cis	0.004	0.25
4-$CH_2OCH_2CH_3$	H	± cis	0.5	
4-CHOCH$_3$ $\vert$ CH$_3$	H	± cis	>5	
4-CH_2NMe_2	H	± cis	0.25	
4-$CH_2SC_2H_5$	H	± cis	0.1	
4-OC_6H_5	H	± cis	0.04	~0.7
4-$SiMe_3$	H	± cis	0.05	0.25
4-F	H	± cis	0.02	
4-NO_2	H	± cis	>5	

[a] Filterpaper-test

Table 40. Absolute toxicities LD_{50} of some chrysanthemates against fly and locust [64]

R	Stereochemistry	LD_{50}	
		Housefly ng/fly	Locust ng/kg
	1Rtrans; S (Pyrethrin I)	140	11
	1Rtrans; S (Esbiol)	110	11
	1R/Strans; S	1200	
	1Rtrans; RS	670	
	± cis/trans; R, S	990	
	± cis/trans	100	
	± cis/trans	150	
	± cis/trans Resmethrin	340	
	1Rtrans	25	

Table 40. (continued)

R	Stereochemistry	LD$_{50}$	
		Housefly ng/fly	Locust ng/kg
	(±) cis/trans Tetramethrin	800	37
	± cis/trans	9000	
	± cis/trans	inactive	
	(±) cis/trans Phenothrin	85	8
	(±) cis/trans	48	

2.2.1.2.2 Structurally Rigidized Alcohols

Table 41. Insecticidal activity of close analogous of pyrethrin I [151]

1Rtrans
R

	Fly			Cockroach
	Topical µg/g		Contact µg/cm^3	Contact µg/cm^3
	with PBO	syner-gism factor	LD$_{50}$ (18 h) KD$_{50}$ (15 min)	LD$_{50}$ (24 h)
pyrethrin I	(S)	26 0.08	320	>0.3 0.13	
jasmolin I	(S)	95 0.68	140	0.23 >0.6	
cinerin I	(S)	85 0.27	320	0.22 0.14	
	(RS)	260 0.7	380	>0.6 >0.6	

Table 41. (continued)

1Rtrans R			Fly					Cockroach
			Topical µg/g			Contact µg/cm³		Contact µg/cm³
				with PBO	syner-gism factor	LD_{50} (18 h)	KD_{50} (15 min)	LD_{50} (24 h)
bioallethrin		(RS)	29	0.68	40	0.04	0.1	
		(RS)	12	0.60	20	0.03	0.08	0.33
			50	0.55	90	0.13	0.3	0.20
			inactive [151a)					
			active [152]					
			more active than Alle-thrin [153]					
			inactive [71]					

Table 42. Insecticidal activity of some rigidized 3-phenoxyben-zylpyrethroids [154]

Alcohol	Rel. toxicity to	
	Housefly	Mustard beetle
	0.3	3
	<0.1	<0.1
	1	<0.1
	1.2	2.8
	<0.1	<0.1
Bioresmethrin	100	100

Table 43. Relative insecticidal potency of dibenzofuranmethanol pyrethroids against *Phaedon. Laphygma, Plutella* [156, 157]

Position	Rel. Tox.	calc. Energy E_{HOMO} of dibenzofuran methanol
1	20 $(LC_{90} 4.10^{-3}\%)$	−10.95
4	8	−10.82
3	1	−10.78
2	Inactive	−10.67

Table 44. Relative toxicity of dihydrobenzofuranol-chrysanthemates against cockroach [158]

CO—R	Rel. tox.		Rel. tox.
Allethrin	100		
	10		160
	120		10
	100		–
	10		
	80		100

Table 45. Relative insecticidal activity of torsionally fixed 3-biphenylmethylester-pyrethroids [159]

Compound	Insect			
	Topical			Foliar application
	Spodoptera	*Epilachna*	*Oncopeltus*	*Acyrthosiphon*
Permethrin dosis per ins.	1 (23 ng)	1 (13 ng)	1 (660 ng)	1 (21 ppm)
(±) cis n = 1	0.01	0.03		0.2
2	0.5	2.3	4.3	4.9
3	0.5	3.0	6.5	77
4	0.06	0.6	1.5	3.6
„ 2	0.7		1.7	
3	1.0	0.8	13	33

Table 45. (continued)

Compound	Topical			Foliar application
	Spodoptera	Epilachna	Oncopeltus	Acyrthosiphon
CH₂— (with ring) (CH₂)ₙ— (pyrrole ring) n = 2	0.8	0.2	2.0	1.3
n = 3	0.2	0.2	1.2	8.5
Cl₂C=CH–cyclopropane–COOCH₂–(tricyclic aromatic) 1Rcis	2.9	2.3	19	10.4
Br₂C=CH–cyclopropane–COO " 1Rcis	1.0	1.3	21	5
(dimethylvinyl cyclopropane)–COO "	0.5	0.5	1.4	
Cl–phenyl–CH(iPr)–COO "	—	0.2	0.5	
Et–O–phenyl–C(cyclopropane with Cl, Cl)–COO "	0.01	0.1		0.8

Table 46. Insect-toxicity of 2-indanyl pyrethroids

Y	Southern armyworm Rel.	ppm	Mexican bean beetle Rel.	ppm	Aphid Rel.	ppm	Boll-weevil ppm	Spider-mite ppm	Lit.
X = Cl									
1Rcis; (R', S') (phenyl)	1	24	1	5	1		13	7.8	[160]
o-phenyl	0.2		0.1		0.03				
thiophene-S			0.7		0.06				
Br	0.25		0.2		0.03				
CH₃	–		–		–				
X = CF₃ ± cis, (R'S') (phenyl)	1.3	18	2	2.6	10		4	3.8	
1Rcis Z; S' (phenyl)		0.2		0.8		5.6	2.2	1.5	
cis Permethrin		1.1		7.2		240	12	320	

[161]

2.2.2 Non-ester Pyrethroids

Table 47. Toxicity of oxime-type pyrethroids to different insects [90, 162, 163]

X	R	syn/anti (Z/E)	ppm (% mortality)				(topical µg/ins.)	
			Musca	*Phaedon*	*Spodop-tera*	*Aëdes*	*Helio-this v. larv.*	*Musca adults*
4–Cl		0 : 1	10 (75)	10 (90)	100 (100)	1 (100)	0.17	0.2
		1 : 0	500 (50)	500 (100)	500 (90)	1 (100)	13	8

Table 47. (continued)

X	R	syn/anti (Z/E)	ppm (% mortality)				(topical µg/ins.)	
			Musca	*Phaedon*	*Spodop-tera*	*Aëdes*	*Helio-this v.* larv.	*Musca* adults
	△	mix. 0:1 1:0	10 (40)	10 (70)	100 (100)	1 (100)	0.083 1.1	0.07 0.64
	⟩	mix.	500 (80)		500 (100)			
	⟩o	1:0	500 (90)	100 (95)	100 (65)			
4– ⟨	⟨	1:0	500 (100)	10 (60)	100 (25)	10 (100)		
4–OEt	⟨	1:0	100 (100)	10 (35)	100 (100)	1 (100)		
3–Cl	⟨	mix.	100 (100)	10 (70)	10 (30)	1 (100)		
3–CF₃	⟨	mix.	500 (100)	10 (80)	500 (90)	1 (70)		

Table 48. Toxicity of oxime-type pyrethroids against *Bombyx mori* [22]

X	Syn/anti	LC_{50} (ppm)
2-Cl	1:1	2100
4-OCH₃	1:1	5
3,4-CH₃	1:1	16
3,4-O−CH₂−O	1:1	14
3-Cl, 4-CH₃	1:1	1.8
3,4-Cl₂	1:1	1.4
4-CH₃	3:1	2
4-Cl	3:4	1.7
H	6:5	18

Table 49. Toxicity of oxime-type pyrethroids [66]

R_1	R_2	R_3	Stereo-isomer	LD$_{50}$ mg/kg (topical)		
				Blowfly	Cockroach	Budworm
Cl	CHCl$_2$		E (cis)	25	>73	11
	CHCl$_2$	,,	Z (trans)	1.2	8.5	1.1
EtO	CHCl$_2$	,,	E/Z 90 : 10	1.2	>18	>8
		,,	Z[a]	0.4	4.5	2.3
EtO	CHCl$_2$		E (cis)	1.4	>9	1
		,,	Z (trans)	0.7	4.5	0.4
Cl	CHCl$_2$		E (cis)	12	27	14
Cl	CHCl$_2$	,,	Z (trans)[b]	83	>73	>32
EtO	CHCl$_2$	,,	E	0.34	10	3
EtO	CHCl$_2$	,,	E/Z 20 : 80	1.6	10	3.5
EtO	CCl$_3$	,,	E	0.73	32	2.7
EtO	CF$_3$	,,	E (cis)	3.3	>36	23
EtO	CF$_3$	,,	Z (trans)	34		
Cl		,,	E (trans)[c]	9.4	17	10
Cl	,,		Z (cis)	180	>73	>32

a   trans   b   cis   c   trans

Table 50. Insect toxicity of some ketone-type-pyrethroids [164a]

	Musca	Locusta	Dysdercus	Droso-phila	Aëdes	Tetra-nychus
	LD$_{50}$ μg/ins.	LD$_{50}$ μg/ins.	LD$_{50}$ μg/ins.	LC$_{50}$ mg/g	LC$_{50}$ mg/l	at 0.1%
	0.023	0.4	0.03	20	0.006	—

Table 50. (continued)

	Musca LD$_{50}$ µg/ins.	Locusta LD$_{50}$ µg/ins.	Dysdercus LD$_{50}$ µg/ins.	Droso-phila LC$_{50}$ mg/g	Aëdes LC$_{50}$ mg/l	Tetra-nychus at 0.1%
	0.27	1.4	0.3	130	0.8	+
,,	0.21	0.2	0.1	140	0.4	+
,,	10	> 100	> 10	830	> 10	−
	1.4	12.6	12	95	> 10	−
,,	0.18	15	0.8	95	0.2	−
,,	0.27	12	0.1	33	0.5	+
,,	0.25	8	0.6	21	2	−
	> 10	> 100	> 10	> 1000	> 10	−
	> 16	> 100	> 10	> 1000	> 10	−
	0.4	> 100	0.2	200	> 10	−
,,	3	> 100	> 100	14	0.5	−

Table 51. Influence of four Important Substituents in Nonester-Pyrethroids on Relative Insect-Toxicity [165]

Relative Potencies: (Bioresmethrin ≡ 100)

X	Y	Z	R	Musca	Phaedon	Fluorine bonus Musca	Fluorine bonus Phaedon	Ethoxy bonus Musca	Ethoxy bonus Phaedon
Cl	CH=CH	H	H	54	12	1.3	6.6		
„	trans	F	H	70	79			1.4	13
EtO	CH=CH	H	H	52	58	1.9	2.8		
„	trans	F	H	100	160				
Cl	CH$_2$CH$_2$	H	H	6	9	3.8	2.6		
„	„	F	H	23	23			0.9	2.5
EtO	CH$_2$CH$_2$	H	H	11	9	1.9	6.3		
„	„	F	H	21	57				
Cl	CH$_2$O	H	H	26	5	0.8	8.6		
„	CH$_2$O	F	H	20	43			1.1	1
EtO	CH$_2$O	H	H	22	5	1.3	8.6		
„	„	F	H	29	43				
EtO	CH=CH trans	H	F	110	130	0.6	5		
	CH=CH trans	F	F	70	26				

Table 52. Variation of Activity with Relative Stereochemistry of Substituents in 1-Phenyl-1-alkoxymethylcyclopropanes [166]

X	Y	Z	LD$_{50}$ mg/kg Lucilia cuprina	Blattella germ.	Heliothis punctigera
H	H	H	5.7	2.8	2.7
H	F	H	3.5	5.7	3.1
F	H	H	2.3	2.0	0.8
F	F	H	1.7	1.9	0.3
F	F	F	1.7	0.9	0.2
H	Cl	H	6.3	48	3.7
Cl	H	H	4.0	6.3	0.1
Cl	Cl	H	11.3	71.8	2.4

Table 53. Relative Activity of Nonester-Pyrethroids [151a] (Resmethrin ≡ 100)

R	A	B	X	Y	*Musca*	*Phaedon*
Cl	(CH3)2C<	CO	O	H		inactive
Cl	,,	−HC=CH− trans		H	6	6
Cl	,,	cis		H		inactive
Cl	,,	−C≡C−		H		inactive
Cl	▽	HC=CH trans		H	54	12
Cl	,,	,,		F	70	80
Cl	,,	−CH₂−CH₂		H	6	9
CH₃O	,,	HC=CH trans		H	11	5
CH₃O	,,	,,		F	52	58
C₂H₅O	,,	trans		F	100	160
C₂H₅O	,,	−CH₂−CH₂−		F		less active
C₂H₅O	,,	−(CH₂)₃		F		inactive

Table 54. Insect Mortality (%) after Foliar Application (10 ppm) [62] Arylalkane-Types of Pyrethroides

X	Y	Tobacco Cutworm	Brown plant-hopper
H	H	—	—
H	F	100	85
EtO	H	75	25
	F	100	95
F₂CHO	H	80	20
	F	100	90
CH₃O	F	10 times less active than EtO	

Table 55. Comparison of two Arylalkylether-Types of Pyrethroids [62]

Animal			Toxicity (LD$_{50}$ Topical µg/g)		Fluorine bonus
			MTI 500 X = H	MTI 800 X = F	
Nilaparvata			1.95	0.61	3.2
Laodelphax			0.5	0.2	2.5
Nephotettix	}	Rice pests	0.4	0.27	1.4
Chilo			7.2	0.6	12
Lissorhoptrus			7.5	2.5	3
Callosobruchus	}		0.1	0.07	1.7
Plutella		Vegetable pest	42	3.4	13
Spodoptera			0.1	0.11	1
Blattella	}		0.74	0.14	5.3
Musca		Household pest	0.15	0.05	3
Culex			0.02	0.005	4
Mouse			LD$_{50}$ i.p. > 30 g/kg (!)		
Carp			TLM 48: 5 ppm 40 ppm		0.12

Table 56. Activity of Simple Benzylether-Types of Pyrethroids against Rice Pests [167]

				LC$_{50}$ (ppm)		
				Neph. c.	*Nilap. l.*	*Laodelph.*
X	Y	Z	W			
CH	O	H	H			
CH	O	F	H	1.3	26	33
CH	O	F	F	2.0	39	50
CH	O	F	Cl	43	7.6	
CH	O	F	CH$_3$	8.7	6.9	
N	O	H	H	5.4	3.4	15
N	O	F	H	4.5		
N	O'	F	F	1.3		
N	O	F	Br	6.2		
CH	NH	F	H	1.5	1.5	
CH	NH	F	CH$_3$	30	18	

X	Y	Z	W			
CH	O	F	H	11	6.7	15
N	O	H	H	66	50	
CH	NH	F	H	21	30	

Table 56. (continued)

| | | | | | LC$_{50}$ (ppm) | | |
					Neph. c.	Nilap. l.	Laodelph.
A	X	Y	Z	W			
CH$_3$	CH	O	F	H	1.3	26	33
Cl	CH	O	F	H	2.4	22	18
Cl	N	O	H	H	15	39	90
Br	N	O	H	H	32		
CH$_3$O	N	O	H	H	99		
Cl	CH	NH	F	H	45	18	
	Malathion				500		

Table 57. Effects of Benzylether Types of Pyrethroids against Brown Planthopper [168]

	Appl. rate (ppm)	Mortality %
x = 3 Cl	100	40
	20	3
3,4 OCH$_2$O	100	75
	20	10
4 Cl	100	100
	20	100
4 CH$_3$	100	100
	20	20

	100	100
	20	50

	100	100
	20	15

(Toxicity partly due to metabolic degradation to fluoroacetic acid?)

Table 58. Effects of Carbamate-Types of Pyrethroids against Flies [169, 169a]

	LD$_{50}$ µg/g	
	Unsynergized	Synergized with PBO

| | 30 | 0.6 |

,,

Table 58. (continued)

		LD$_{50}$ µg/g	
		Unsynergized	Synergized with PBO
R-isomer	,,	46	1
S-isomer	,,	>600	28
(structure)	,,		35
(structure)	,,	24	1.5
(structure)	,,	55	5
(structure)	,,	>600	170
(structure)	,,	520	8
(structure)	,,	270	6
(structure)	,,	>600	500
(structure)		1.2	0.12

2.2.3 Structure and Speed of Action

Table 59. Relative Speed of Action KT$_{50}$ (knock-down) of some Pyrethroids against Household Pests [170]

		Musca	Culex	Blattella
(±) cis/trans	Tetramethrin	1[a]	1[b]	1[c]
	Bioallethrin	0.9	0.35	1.1
1 R trans	(structure)	1.2		4.3

Table 59. (continued)

			Musca	Culex	Blattella
,,	,,		3.9		4.3
,,		,,	6.2		4.5
		,,	2.5		
	,,		4.1		2.5
	,,		1.8	1.9	4.8
			2.2		2.7

KT_{50} (min): Time for knockdown for 50% of insects at 6.25 mg/l
[a] KT_{50} 8.9 min; [b] 2.7 min; [c] 5.3 min

Table 60. Relative Speed of Action (KT) and Relative Lethal Toxicity of Synthetic Pyrethroids against Locusts [171]

	Log. rel. KT	Rel. Tox.
Deltamethrin	0.43	3200
1 R trans	0.4	190
1 R trans, S ,,	0.37	100
(±) cis/trans	0.36	260
(±) Fenvalerate	0.32	
K'Othrin	0.06	98

Table 60. (continued)

	Log. rel. KT	Rel. Tox.
Bioresmethrin	0	100
1 R cis Bioresmethrin		96
1 R cis	-0.075	760
Cypermethrin	0.025	

Table 61. Relative Knock-Down-Rates of some 1 R cis-Z-Norpyrethric Esters against *Musca d.* [107]

R^1	R^2		
CH_3	630	106	74
C_2H_5	280		1000
CH_2CF_3	180	420	580
iC_3H_7	70	280	1330
nC_3H_7	140	160	920
Bioallethrin	150	100	100

Table 62. Knock-Down Effect KT_{50}) of Pyrethrin I-Analogues [172][a]

Compound	Stereochemistry	Mosquitoes KT_{50} sec.	Mortality %	Houseflies KT_{50} sec.	Mortality %
	1 R t; S	320	73	> 600	
		480			
„	± c/t; RS	300	83	500	30

Table 62. (continued)

Compound	Stereochemistry	Mosquitoes		Houseflies	
		KT_{50} sec.	Mortality %	KT_{50} sec.	Mortality %
	±c/t; RS	120	90	130	40
	" ± c/t; RS	85	100	83	60
	" ±c; RS	81	100	80	50
	" ±t; RS	89	97	86	67
	" 1 R cis, RS	100	100	95	50
	" 1 R trans, RS	44	100	40	80

[a] oil spray $0.5 \text{ mg}/0.3 \text{ m}^3$

Table 63. Relative Knock-Down-Rate of Racemic Fenpropana-te-Analogues against *Musca* [173]

R	Rel. k.d.	R	Rel. k.d.
	1		3.0
	2.4		3.2
	2.7		3.3

Table 63. (continued)

R	Rel. k.d.	R	Rel. k.d.
	2.9		3.5

2.3 Structure-Activity-Analysis

2.3.1 Stereochemical Considerations

The vast amount of biological activity data on pyrethroids published in the last 10 years is valuable for the elucidation of the molecular origin of insecticidal action, to get information of the chemical surrounding of the binding site, the factors influencing pharmacokinetic and metabolism, the shape of active isomers and even for predicting novel molecules for action at the pyrethroid binding site. Last not least, it is an ideal testground for approaches in complex quantitative structure-activity-analysis, as presented in the publications by Lee, Hopfinger, Ford, Fujita, Matsuo, Plummer and their co-workers. Since the early investigations on the insecticidal action of pyrethrum by Staudinger and Fujitani 80 years ago it was clear, that only the intact ester compound is active. There is no single 'toxophore' in pyrethroids. Later on it was established that the insecticidal action is extremely stereospecific, that only certain stereoisomers out of a number of possible ones are active. The absolute configurations of acid or alcohol components necessary for activity are shown in Table 64.

Table 64. Necessary absolute configurations of pyrethroid acids

Table 64. (continued)

Table 65. Preferred or necessary absolute configurations of pyrethroid alcohols and non-ester-pyrethroids

Table 65. (continued)

Both cis or trans configurated cyclopropanes give active esters. It depends on the insect species, which is the more active one (Tables 3, 66). In the case of the very active m-phenoxybenzyl esters in the majority of cases the cis isomer is more active, either intrinsically or due to selective and faster metabolization of the trans-isomer. In certain cases, the trans isomer is more active than the cis ester, as in the chrysanthemates pyrethrin, allethrin, phthalthrin and resmethrin [179b]. Trans permethrin is the 8 times more active isomer against the xylophagus insect *Hylotropas bajulus* [180]. The trans-isomer of certain non-phenoxybenzyl esters are more active against housefly [180a], however, in a higher fluorinated series the cis isomers are claimed to be the ones with higher activity [181].

In case of fenvalerate, the optically pure single isomeric ester of the S acid and α R alcohol is 70 times less active against housefly but 9 times less active against mosquito than the S, α S-isomer [101]. Similarly, 1 R trans α R cypermethrin is 95 times less active to adult houseflies, but only 5 times less active to mosquito larvae than the 1 R trans αS isomer [101]. The ester of R-fenvaleric acid are essentially inactive. See also p. 112.

2.3.2 Conformational Analysis of Pyrethroids

X-ray analysis was important for establishing absolute configuration of the active isomers, but was of little help for understanding the relation of molecular shape to biological activity. According to X-ray analysis, the structures in the crystal-lattice are as shown in the following computer drawn stereoformulas (p. 68–75):

Deltamethrin [182]. Insecticidally active isomer 1R-cis, α-S (Fig. 1):

Cypermethrin [183]. Insectidally inactive isomer 1R-trans, α-R (Fig. 2):

Permethrin [184]. Insectidally active isomer 1R cis isomer (Fig. 3):

Descyanodeltamethrin [184]. Insectidally active isomer 1R cis (Fig. 4):

Cyfluthrin [185]. Insectidally inactive cis-racemate I (= 1R cis α R/1S cis α S): (Fig. 5)

Cyfluthrin. Insectidally active cis racemate II (1R cis α S/1S cis α R): (Fig. 6)

Cyfluthrin. Insectidally inactive trans racemate III (1R trans α R/1S trans α S): (Fig. 7)

Cyfluthrin. Insectidally active trans racemate IV (1R trans α S/1S trans α R): (Fig. 8):

Table 66. Insect-toxicological relevance of the three asymmetric centres in Cyfluthrin for different insect species[a] [174][b]

Stereoisomeric compounds tested		Comparison	Ratio of approximated relative toxicities LC$_{50}$ B/LC$_{50}$ A							
A	B		Dr.	Ph.	Plu.	He. arm.	He. vi.	Lap.	Myz.	Tetr.
A versus B										
1R cis α(RS)	1S cis α(RS)	1R cis/1S cis	10	500	50	70	35	50	100	500
1R trans α(RS)	1S trans α(RS)	1R trans/1S trans	100	100	50	3	3	100	100	10
1R cis α(S)	1R cis α(R)	αS/αR		50	15			50	270	70
1R cis α(RS)	1R trans α(RS)	1R cis/1R trans	1	5	1			1	1	5
1S cis α(RS)	1S trans α(RS)	1S cis/1S trans	1	1	1			1	1	1

[a] Dr.   Drosophila
Ph.   Phaedon
Plu.   Plutella
He. arm. Heliothis armigera
He. vic. Heliothis virescens
Lap.   Laphygma
Myz.   Myzus
Tetr.   Tetranychus

[b] Relative values, the inactive compounds were not 100% isomerically pure (~98%)

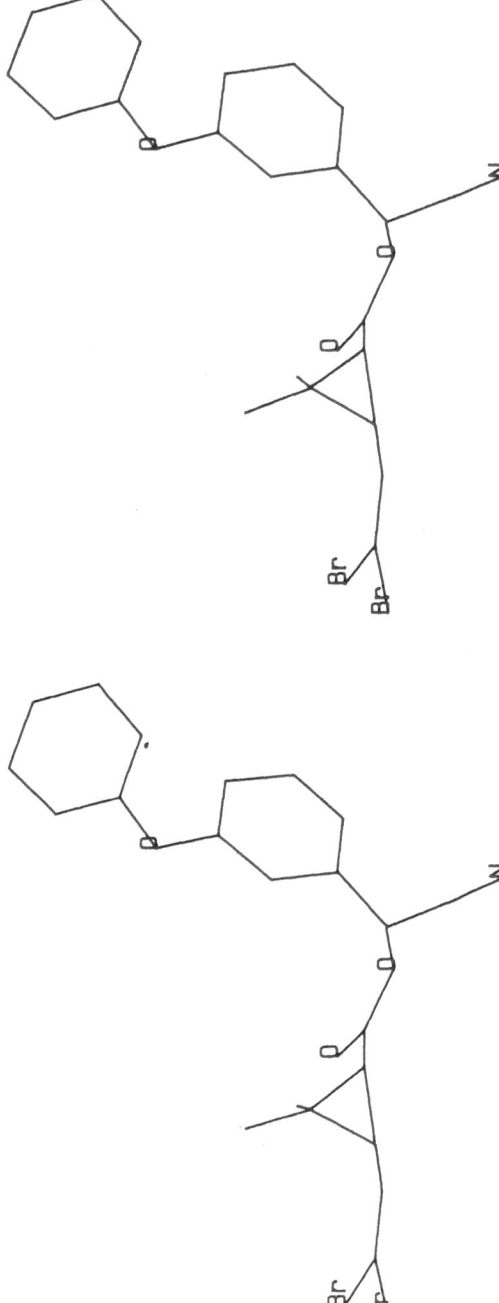

Fig. 1. Conformation of deltamethrin 1R cis, α S single isomer in the crystall lattice. Extended conformation, the phenoxy group is as far away as possible from the dibromovinyl group. The cyano group and dibromovinyl group are on the same side of the plan of the cyanopropane ring.

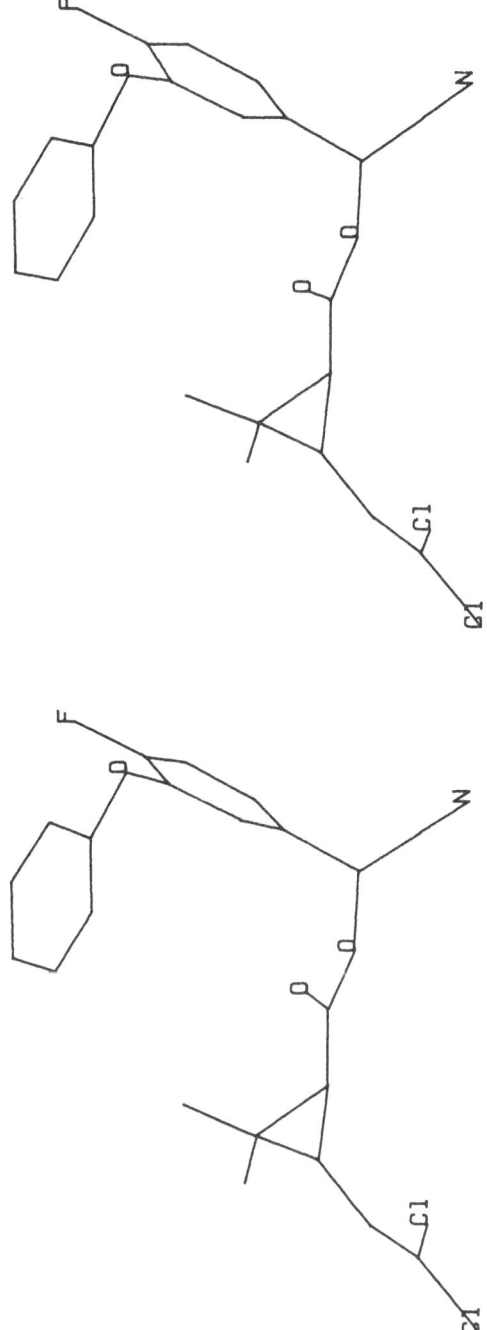

Fig. 2. Conformation of cypermethrin 1R trans, α S single isomer in the crystall lattice. Folded conformation, phenoxy group and dichlorovinyl group point in the same direction, CN group and CO group perpendicular to each other point in the opposite direction.

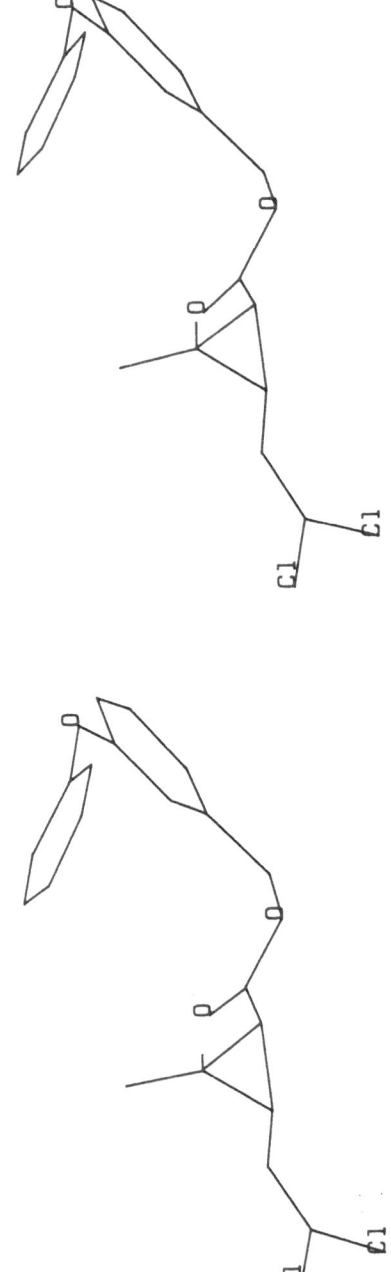

Fig. 3. Conformation of permethrin 1R cis single isomer in the crystall lattice. Filded conformation, phenoxy group in the vicinity of the CH$_3$ group, which is trans to the ester moiety.

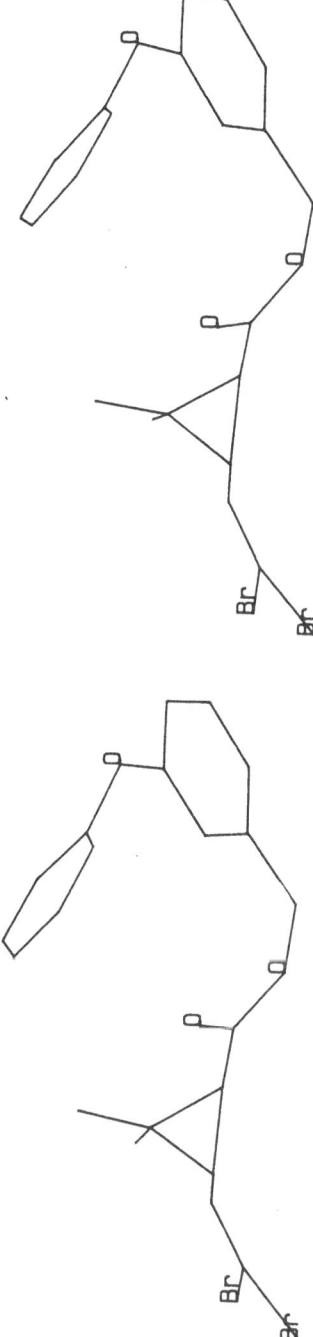

Fig. 4. Conformational of descyano-deltamethrin 1R cis single isomer in the crystall lattice. Folded conformation, phenoxy group in the vicinity of the CH_3 group which is trans to the ester moiety, the same conformation as 1R cis permethrin.

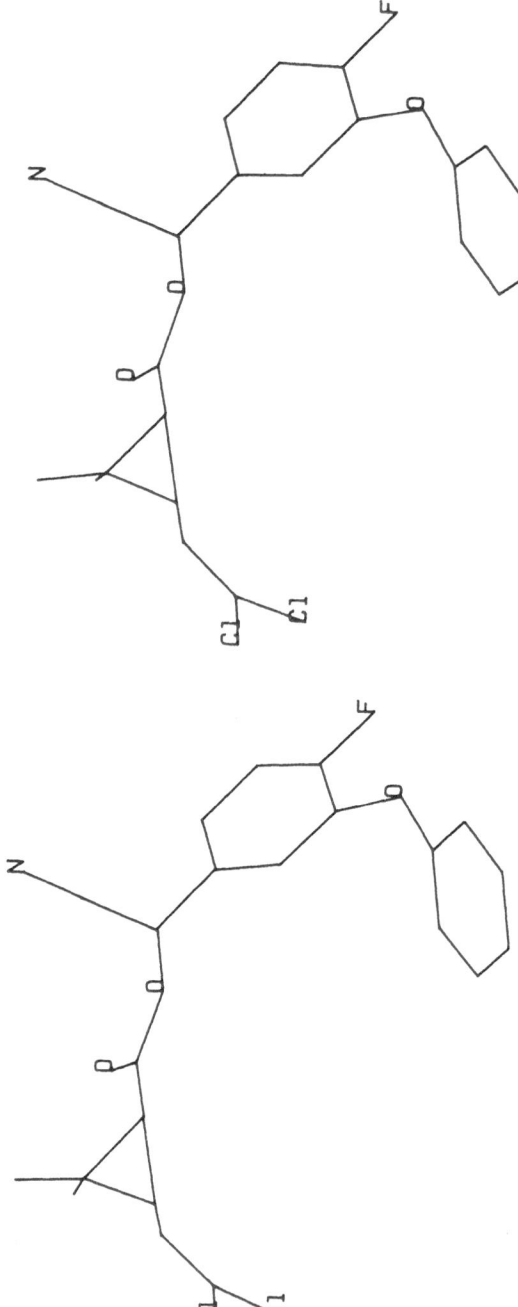

Fig. 5. Conformation of Cyfluthrin 1R cis, α R/1S cis, α S racemate I in the crystall lattice. Folded conformation, very similar to the inactive trans-cypermethrin in Fig. 4.

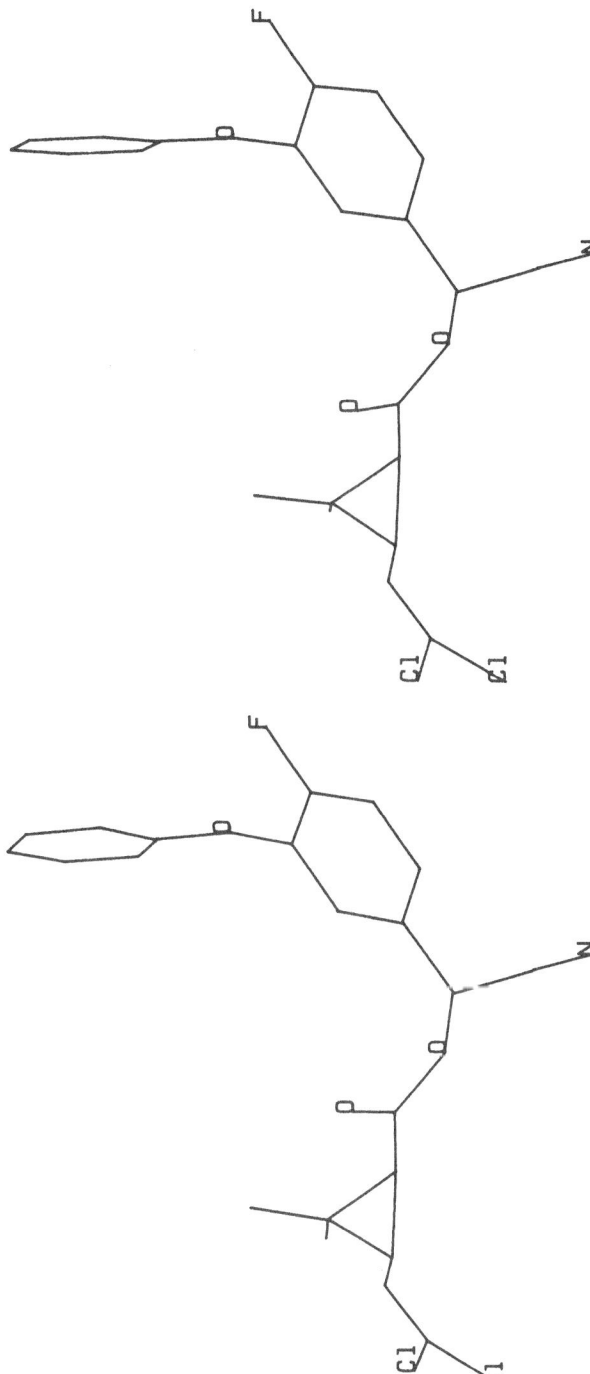

Fig. 6. Conformation of cyfluthrin 1R cis α S/1S cis α R racemate II in the crystall lattice. Folded conformation, phenoxy group and methyl group, cis to the ester moiety, both point in the same direction.

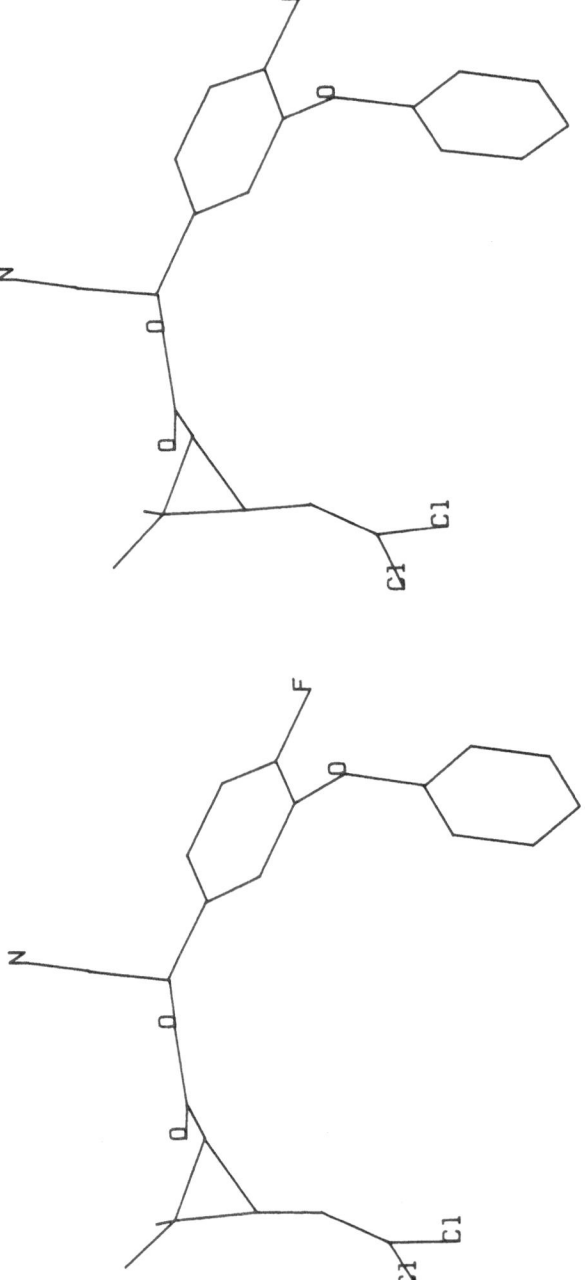

Fig. 7. Conformation of cyfluthrin 1R trans α R/1S trans α S racemate III in the crystall lattice. Folded conformation, very similar to conformation in racemate I.

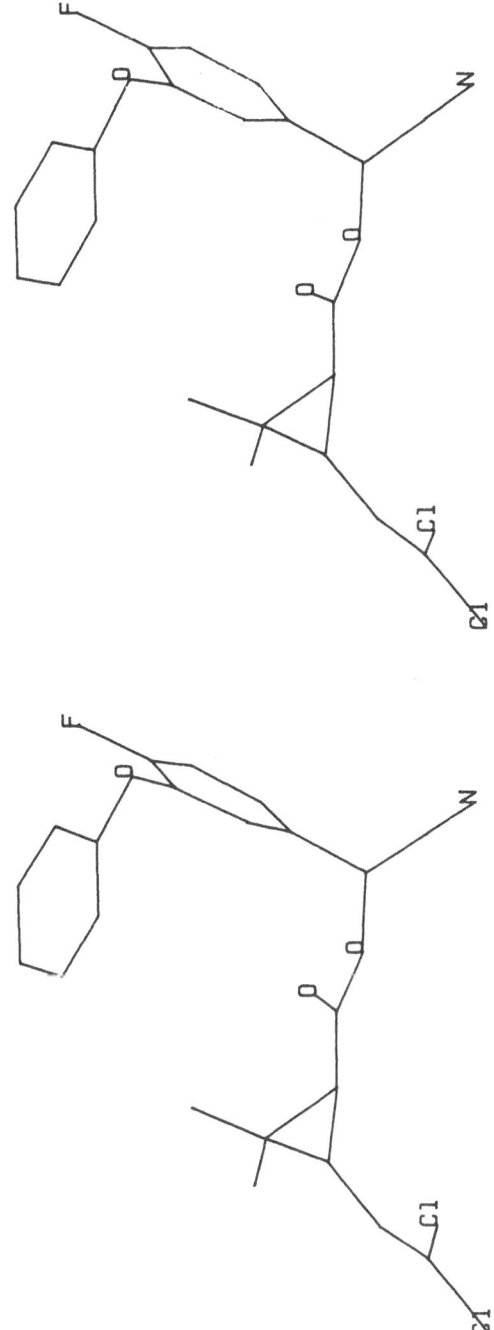

Fig. 8. Conformation of cyfluthrin 1R trans α S/1S trans α R racemate IV in the crystall lattice. Folded conformation, phenoxy group and methyl group, cis to ester moiety, both point in the same direction.

4 Shape of Active Conformations of Pyrethroids in the Target Site Deduced from QSAR

A flexible molecule like deltamethrin has 7 torsional axes, hence a great number of possible conformations. Reasonable assumptions concerning the conformation of ester-pyrethroids would put the ester-group

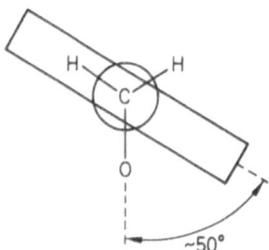

in a trans-coplanar situation, and assign for such lipophilic molecules in a lipophilic environment an extended conformation [186]. This educated guess reduces the number of likely conformations. A more detailed analysis revealed, that in addition the phenoxy moiety is also conformationally rigid and also the substituent at the cyclopropane ring, so that only two modes of rotation remain, which are of importance [187, 188]. The optimal angle of this meta-substituted phenyl plane with the plane of the $O-CH_2-C$ bond was found to be about 45 to 50° [189].

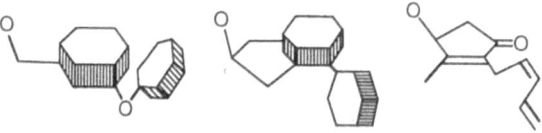

The orientation of the aromatic substituent at the benzyl moiety or cognate situation in pyrethroid esters in the active conformation is out of plane of the first phenyl ring (Fig. 9).

Fig. 9. Geometrical orientation of the phenyl plane for optimal insecticidal activity of benzyl ester pyrethroids

Rigidized planar structures are inactive.

As seen in the X-ray structure, active and inactive isomers may accommodate either a folded or extended conformation in the crystall lattice.

NMR analysis of fenvalerate revealed a preferred conformation in carbon tetrachloride in which the CN group and the chlorophenyl group are aligned parallel to each other in close vicinity [142]. In trifluoroethanol however, a folded shape of the molecule is prevalent [187], putting the two aromatic systems in close proximity. The same 'horseshoe' conformation of a pyrethroid was observed in a lipid bilayer [192].

A more sophisticated probabilistic automatic pattern analysis [190] of a set of 35 esters from 10 acids and 24 alcohol components resulted in a common conformation needed for insecticidal action, which explains most of the variation of LC_{50} and knock-down against fly and cockroach. This analysis was based on overall electronic effects, estimated in terms of relative Lewis acidity and basic bonding deduced from tlc-data (obtained from specially prepared tlc-plates), geometric bulk and partitioning coefficients.

The results are shown in the following table of correlations. These principal correlations were often corroborated later on [182]:

Table 67. Correlation of knock down effect and toxicity with electronic and steric properties of some pyrethroids

	LC_{50}	Knock-down
Electronic effects	+	−
lg P	−	+
certain aspects of peculiar conformations (bulk + electronic effects)	+	−

This peculiar folded 'clamp or horseshoe' conformation of the model pyrethroid

trans

corresponds to the X-ray-structure of the *inactive* 1R trans α R-cypermethrin isomer (Fig. 2, p. 69). The importance of certain aspects of the molecule for toxicity is shown in the following figure 10 [190].

The view from A and B shows very significant aspects of the active conformer. The high r-values designate the importance as recognition features of the molecule in this active conformation by the receptor. The aspect from C means the projection of the molecule in the plane, yielding the ordinary graph of the structural formula of the pyrethroid as written by chemists for clarity's sake. Within certain limitations this aspect is quite variable, as shown in the diversity of formulae of patented pyrethroids, but is not important for the recognition by the receptor as seen by the small r-value.

Another conformational analysis of α-substituted phenothrin analogues, connecting toxicities of houseflies with lgP-values and calculated energy differences of possible conformers and stereoisomers, resulted in a correlative structure activity map and in the active conformation during insecticidal action. This is again the 'horseshoe' as previously mentioned, in which the phenoxy-substituent of the alcohol and the isobutenyl moiety of the acid component are in vicinity. The same conformation holds for fenvalerate, where the phenoxy substituent and the chlorophenyl ring are

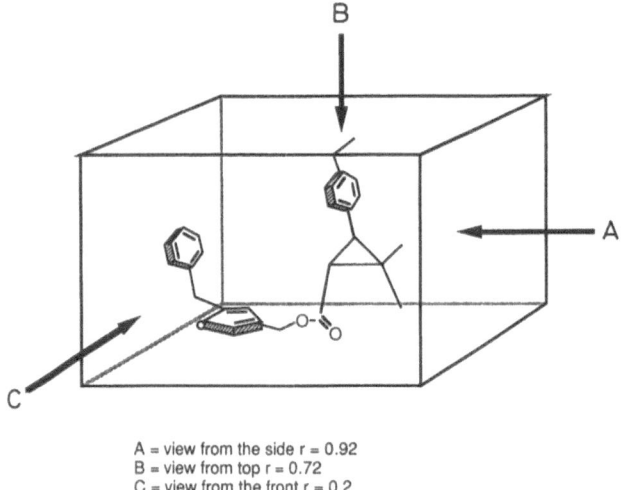

A = view from the side r = 0.92
B = view from top r = 0.72
C = view from the front r = 0.2

Fig. 10. Proposed conformation of a pyrethroid necessary for insecticidal activity and importance (r-values) of faces of the molecule for biological action

parallely oriented [187, 188]. This conformation can be observed in reality by NMR in trifluormethanol solution. Less than 1 kcal/Mol must be expended. It is proposed that this 'clamp' embraces a planar ring structure in the protein binding site (Fig 11).

Multifactorious QSAR, using knock-down of mustard beetle and neurophysiological effects in leech, size and electronic properties of a number of substituted benzyl chrysanthemates, which are not substituted by phenoxy or phenyl, together with the corresponding permethric esters also yielded this active folded conformation, which is relevant for knock-down effect. In this conformation the phenyl ring is near the methyl group, which is trans to the ester-linkage [191] regardless of cis- or trans-conformation in the acid. Here, in contrast to the example before, it is the first

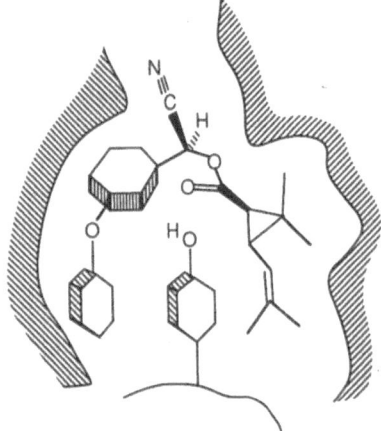

Fig. 11. Modell for binding of a pyrethroid molecule in a binding site of the target membrane bound protein

phenyl ring, which has to adopt a different torsional angle in respect to the plane of the ester-moiety in order to fulfill its clamp function. Another substituent in the benzylic position would prevent this. May be, this is an explanation for the unique α-cyanoeffect, found only in meta-phenoxybenzyl esters. The folded ('horseshoe')-conformation is also adopted in a lipid bilayer near the asymmetric center of the glycerol esters [192]. The active conformation may also explain inactivity in the following isomeric esters [189] (Fig. 12):

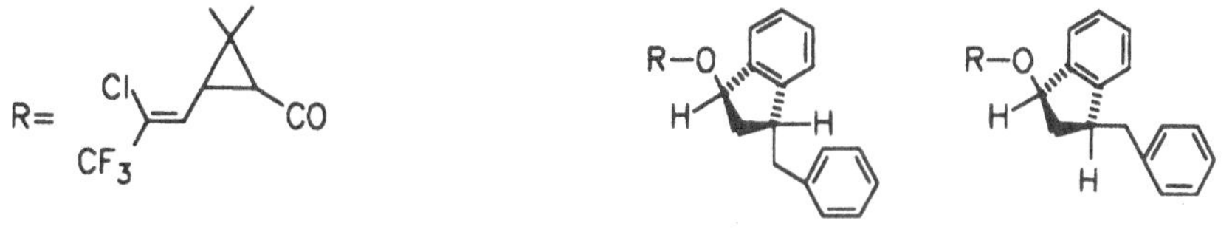

active trans-
isomer

inactive cis-
isomer

Fig. 12. Geometrical orientation of the Π-substituent in two active and inactive pyrethroid isomers

Similarly the inactivity of the cis-isomer of

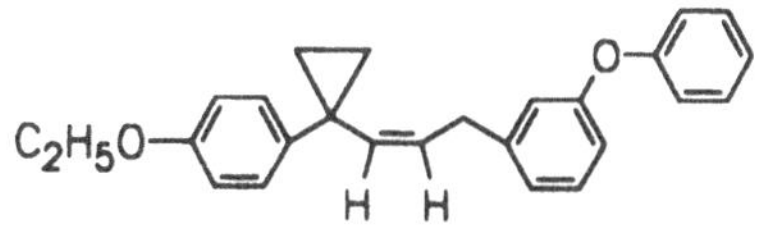

in the achiral pyrethroids of Elliott can be understood. Only the trans-isomer can accommodate the necessary idealized clamp-conformation [90] (Fig. 13):

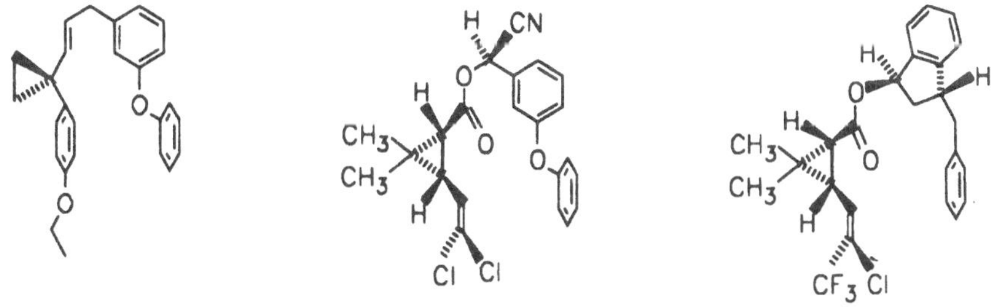

Fig. 13. Application of the clamp/horseshoe model of the bioactive conformation of a pyrethroid for the explanation of biological activity of three diverse pyrethroid structures in relation to their stereochemistry

This assignement of the horseshoe conformation is still tentative and topic of other very recent publications [845, 846]. Certainly this issue will keep being a testground for even more sophisticated multifactorial QSAR, computer modelling and graphics. The clamp model, together with the information on active compounds as represented by the great number of patents on one side and inactive compounds on the other, reveals regions of a pyrethroid molecule which can be modified to retain all or certain activity and which ones are invariable, as shown by a benzyl cyclopropane carboxylate:

x = somewhat variable
y = quite variable
h = hydrophobic, preferentially Π-residue

However, despite the very similar electrophysiological effects concerning Na-channels as the likely primary target of both pyrethroids type I and DDT [56, 193], as shown in many investigations and exemplified by the most diverse structures of this kind of Na-channel effector,

it seems somehow arbitrary to assume an identical side for the same action observed at the protein-lipid-interface or within the protein, where they may intercalate in trans-membrane hydrophobic helices. It may well be that there are several binding sites for competitive, or non-competitive, agonist or antagonist binding for the same or different Na-channel proteins. This will be found out in future studies on the purified and modified receptors and possible subtypes. Some results already suggest different binding sites, e.g. for pentafluorobenzyl- and metaphenoxybenzyloxime-pyrethroids [194], or even for cis- and trans-isomers of the same pyrethroid [195].

2.3.3 Influence of Substituents

The influence of substituent properties of simply substituted benzylchrysanthemates has been investigated by several groups and in several biological systems.

Electrophysiological and toxicological symptoms in synergized cockroach and crayfish could be analyzed separately. The most important contributions to the variation of effects in terms of effectiv concentration c in a given series comes from hydrophobicity (π), sterical bulkiness (V), electronic properties (ϱ, σ) and specific position of the substituent. For example, the correlation-equations for the different biological effects are shown in the next Table 68.

For the highest neuroexcitatory effects, but not neuroblocking properties of pyrethroids, a specific position for the substituent is needed, which is different for the $-NO_2-$ (p > m > o) and the phenoxysubstituent (m > o > p). In addition, the size of the substituent is important. Steric restrictions are highest in the para-position. Electronic effects are only important for the ortho-position. Furthermore, hydrophobicity of the whole molecule is important for neuroblocking effects, while the hydrophobicity of the meta-substituent only contributes to the variation of the neuroexcitation [189], which is correlated with toxicological symptoms in synergized animals [197] only. However, the toxic symptoms are not changed by variation of the hydrophobicity of the molecule. In general, very high activity is obtained only with certain meta-substituents in the phenyl-moiety.

Toxicity against *Spodoptera eridani* is also correlated with the hydrophobicity of meta-substituents, giving, however, two different correlation groups, in addition to stray shots of meta-substituents, having two atoms which separate the phenyl-substituents from the benzyl-moiety [198].

Good results were obtained with π-values and molrefraction of the meta-substituent [199] for explaining variation. Similar results were obtained with substituted α-cyanobenzylchrysanthemates and toxicity to houseflies [189, 200]. In this case, speed of action is also correlated with hydrophobicity, steric bulkiness and stereoelectronic effect of a meta-substituent. Lethal effects are correlated with knock-down in the presence of a synergist, while when it is absent, it is correlated with knock-down and hydrophobicity. Rapid action of trans cyclopropane esters against mustard beetle and in leech ganglia depends on small substituents in meta and para [191] positions, and is enhanced by σ-electron withdrawing inductive groups, and by π-electron-donating groups [201]. The electronic environment of the methylgroup trans to the ester linkage also contributes to the variation of effects. Pyrethroids, whose partitioning coefficient is very sensitive to solvent interactions, tend to act rapidly [191].

In the case of permethrin [201], the meta-phenoxybenzyl-partialstructure explains 25% of the variation of speed of action, while the dichlorovinyl group possessing 50% of the relevant physicochemical properties explains 50%. Electronic properties account for only 8% of the variation.

Interesting work on multivariate techniques for the prediction or explanation of biological effects in terms of dependence on the structure of pyrethroids, particularly involving pharmacodynamic parameters, were published recently by Ford and coworkers [202, 203, 204].

A special case of structure and activity considerations is the action of tralomethrin, (the dibromo adduct of deltamethrin), and tralocythrin (the dibromo adduct of cypermethrin). The two bromine atoms are easily removed in vitro by a number of reducing nucleophiles and in vivo nonenzymically by glutathione [205]. Thus pretreatment of insects with the thiol reagent ethylmaleiimid gave protection against the insecticidal action of tralomethrin [206].

Therefore, there is strong evidence, that these compounds act as pro-pyrethroids [207a]. The corresponding tetrachloro esters are inactive [208].

However, the two enantiomers of tralomethrin and deltamethrin behave differently at very low concentrations. The S'-Enantiomer of Tralomethrin at about 10^{-13} M

Table 68. Correlation-equation of neurophysiological, toxicological and insecticidal effects on cockroach[a] and crayfish[b] [196]

1R trans

Biol. effect	Substitution	$lg\frac{1}{c} = p\sigma + a\pi - b\pi^2 + c\,\Delta V - d\,\Delta V^2 + C$	n	s	r
Neuroexcitation	Ortho	$lg\frac{1}{c} = 5.50 - 1.01\sigma - 0.30\pi^2 + 0.40\,\Delta V^{a)}$	14	0.37	0.878
	Meta	$= 5.45 \qquad\qquad\quad + 65\,\Delta V - 0.07\,\Delta V^{2\,a)}$	17	0.325	0.859
		$= 5.43 \qquad\qquad\quad + 0.88\,\Delta V - 0.09\,\Delta V^{2\,b)}$	16	0.403	0.877
	Para	$= 5.77 \qquad - 0.34\pi + 1.13\,\Delta V - 0.26\,\Delta V^{2\,a)}$	18	0.398	0.933
		$= 5.85 \qquad - 0.41\pi + 1.40\,\Delta V - 0.32\,\Delta V^{2\,b)}$	16	0.377	0.919
Neuroblock	Ortho				
	Meta	$lg\frac{1}{c} = 3.97 \qquad - 0.28\pi^2 + 0.37\,\Delta V$	20	0.284	0.812
	Para				
Convulsion	Ortho	$lg\frac{1}{c} = 6.55 \qquad\qquad\qquad\quad + 0.33\,\Delta V$	11	0.319	0.832
	Meta	$= 6.32 \qquad\qquad\qquad\quad + 0.58\,\Delta V$	17	0.48	0.90
	Para	$= 6.65 \qquad\qquad\qquad\quad + 0.93\,\Delta V - 0.20\,\Delta V^2$	14	0.37	0.78
Lethal effect	Ortho	$lg\frac{1}{c} = 6.08 - 1.25\sigma \qquad\qquad + 0.37\,\Delta V$	10	0.367	0.925
	Meta	$= 5.80 \qquad\qquad\qquad\quad + 0.61\,\Delta V$	15	0.472	0.922
	Para	$= 6.4 \qquad\qquad\qquad\quad\; + 0.71\,\Delta V - 0.13\,\Delta V^2$	13	0.467	0.572

caused the first symptoms in the free walking cockroach twice as fast as deltamethrin did [209]. In addition, HPLC-analysis of nerve membrane incubated and excited by tralomethrin did not detect deltamethrin [210, 211].

There seems to be an intrinsic neurotoxicological activity of tralomethrin in its own right in addition to the demonstrated degradation to deltamethrin, but it is open, whether this is of toxicological relevance (see p. 89).

B. Biological Action of Pyrethroids Against Living Organisms

3 Action of Pyrethroids Against Arthropod Pests

3.1 Action of Pyrethroids on Living Insects

There are several ways for a pyrethroid to enter the body of an insect to exert its lethal effect.

a) Nonstereospecific [212] and rapid [213] penetration through the cuticula of the body or the feet[1], followed by uptake by hemolymphal carrier proteins, such as lipophorin, which also binds favorably other lipophilic xenobiotics including permethrin [214]. The distribution of radiolabeled bromethrin enantiomers throughout the nervous system is also essentially nonstereospecific [216], although the cis-isomer is distributed faster. While the actual central-nervous-system is tightly protected by the dense sheath of cells of the perineurium (like a blood-brain-barrier), neurosecretory cells are a direct neural-hemolymphal interface [216].

b) After penetration of the cuticula a pyrethroid can diffuse along the epidermic cells, using up energy. According to reference [215] this is the main route of distribution to the CNS site, based on observations, that intrahemolymphal application of a topical LD_{50} was much less effective. Therefore, hemolymphal transport should be unimportant.

c) Another direct way for a pyrethroid molecule to reach the nerves is via the vapor phase together with the metabolic oxygen supply through the dense network of tracheae, which leads straight through the finest tracheal ramifications to the CNS [216], bypassing the perineurium.

d) The nervous system of insects is in direct contact with the environment, namely the sensory organs of the peripherous nervous system outside of the cuticula on head, tarsus and legs.

e) The somewhat longer way by oral uptake, digestion and penetration through the intestines into the hemolymph also plays an important role in the action of the pyrethroid. (Table 69)

However, not only these access routes influence the action of the pyrethroid on insects. The vulnerability of the insects also depends on circadian rhythms [217]. Flies are most susceptible to pyrethroid poisoning shortly before dawn; the LD_{50} drops by the factor of 2, as compared to full day light activity. Blowflies are most susceptible

---

1 Contact-LC_{50} of cyfluthrin on treated surfaces 0.7 mg/m^2 *(Aedes)*, 4.3 mg/m^2 *(Anopheles)* [213a]

Table 69. Oral and contact toxicity of some important pyrethroids against tabacco cutworm *(Prodenia litura)* larvae [218]

	LD_{50} µg/g larva 5th stage		LD_{50} µg/g larva 5th stage		Practical application on 6th stage in wheat	
	Oral	Rel. Tox.	Contact	Rel. Tox.	Rate	%Mortality
Endrin	0.5	100	4.1	100	0.28 kg/ha	75
Chlorpyrifos	2.0	25	4.9	84	0.56	65
Deltamethrin	0.050	1040	0.026	15600		
Flucythrinate	0.33	150	0.33	1230		
Cypermethrin	0.44	110	0.097	4200	0.07	65
Fenpropanate	0.61	84	0.32	1270		
Permethrin	0.64	80	0.29	1400	0.07	80
Fenvalerate	0.81	63	0.40	1020	0.07	80
Cismethrin	0.82	62	0.40	1020		
Bioresmethrin	1.64	31	1.1	370		

Table 70. Temperature dependence of toxicity (LD_{50} ng/larva) of some pyrethroids against *Heliothis vir.* [219]

		7 °C	27 °C	38 °C
Deltamethrin		1.4	3.6	3.1
Permethrin		14	31	108
Fenvalerate		13	20	31
Flucythrinate		7	13	15.2

before sunrise and after sunset. Nerve activity in flies is positively correlated with circadian-variations of LD_{50} and knock-down-time respectively. Penetration-rates do not change with the time of day.

Temperature effects also contribute to the effectivity of certain pyrethroids (Table 70).

Pyrethroids of type I (see later) are more active at lower temperatures, however, not in Tetranychus urticae [219a]. Allethrin at 15 °C is 10 times more toxic to the cockroach or boll weevil in regard to knock-down and mortality than at 32 °C [220, 221]. This is however not a general rule. In *Heliothis vir.* larvae for example the temperature coefficients between 38 °C and 15 °C were as follows [222]:

Phenothrin: − 24,2   Cypermethrin: − 1,8
DDT: − 15,4   Tralomethrin: + 1,5
Permethrin: − 9,0   Fenvalerate: + 2,3
Flucythrinate: − 2,0   Deltamethrin: + 5,5

In resistant larvae this negative temperature coefficient is larger, particularly much larger for flucythrinate and fenvalerate [222a].

In addition, cypermethrin proved to be more effective at 32 °C against flea beetle (EC_{50} 38 g/ha) than at the lower temperatures 20 °C and 10 °C (EC_{50} > 150 g/ha) [223]. The same was found with cockroaches [221]. It seems therefore, that the negative temperature coefficient is more associated with type I pyrethroids, which have the faster action. For potato beetle deltamethrin, alphamethrin and lambdacyhalo-thrin efficacy increased from 15 − 20 °C, then decreased between 20 − 25 °C [224]. The knock-down property of allethrin, cypermethrin, deltamethrin was negatively correlated, while the killing power for the CN-pyrethroids showed a positive coefficient [221].

In colorado beetle, the toxicity of cypermethrin, fenvalerate and flucythrinate and permethrin decreases 3 to 8 fold with increasing temperature from 14 to 30 °C, but from 35 °C on again a trend to higher toxicity was observed [225]. However, at predawn times this negative temperature coefficient in blowfly disappears [217].

Another factor influencing the activity of a pyrethroid is the status of an insect. Certain stages are more vulnerable by pyrethroids. For example the L_5 larvae of *Laphygma frugiperda* are about 25 times less sensitive than the L_2 stage [226] as shown in laboratory tests.

Once the pyrethroid molecule has finally reached the nerves, the local selective metabolism at the nerve-cord could be an important determinant of active isomer concentration [226a]. There is no difference in the metabolism of the inactive isomers of 1Scis- and 1Strans-bromethrin, as found out in in-vivo studies in the LD_{99} range [226a]. The main burden of degradation in insects in many species is of an oxidative nature [227]. Metabolism in insects depends very much on the structure of the pyrethroid and on insect species [231].

On the other side, esterase inhibitors have strong synergistic effects in *Chrysopa carnea, Trichoplusiani, Spodoptera* and *Musca domestica,* but not in *Tribolium castaneum,* for example. Mixed functional oxidase inhibitors are synergists in *Musca domestica* (but not in *Spodoptera* and *Trichoplusiani* [227]), and not for the trans-permethrin and fenvalerate in *Tribolium.* However, cis-permethrin and trans-cyperme-thrin are synergized in the last insect [232]. In most cases the trans-isomer is hydrolyzed faster (eg. in *Spodoptera, Heliothis, Periplaneta, Musca*). On the other hand, *Chrysopa carnea* hydrolyzes the cis-isomer faster [227]. There is no general rule for the detoxifying metabolism in insects.

A quantitative structure-pharmacokinetic analysis allows a resasonable prediction of distribution processes in insects [233]. For more detailed information on qualitative

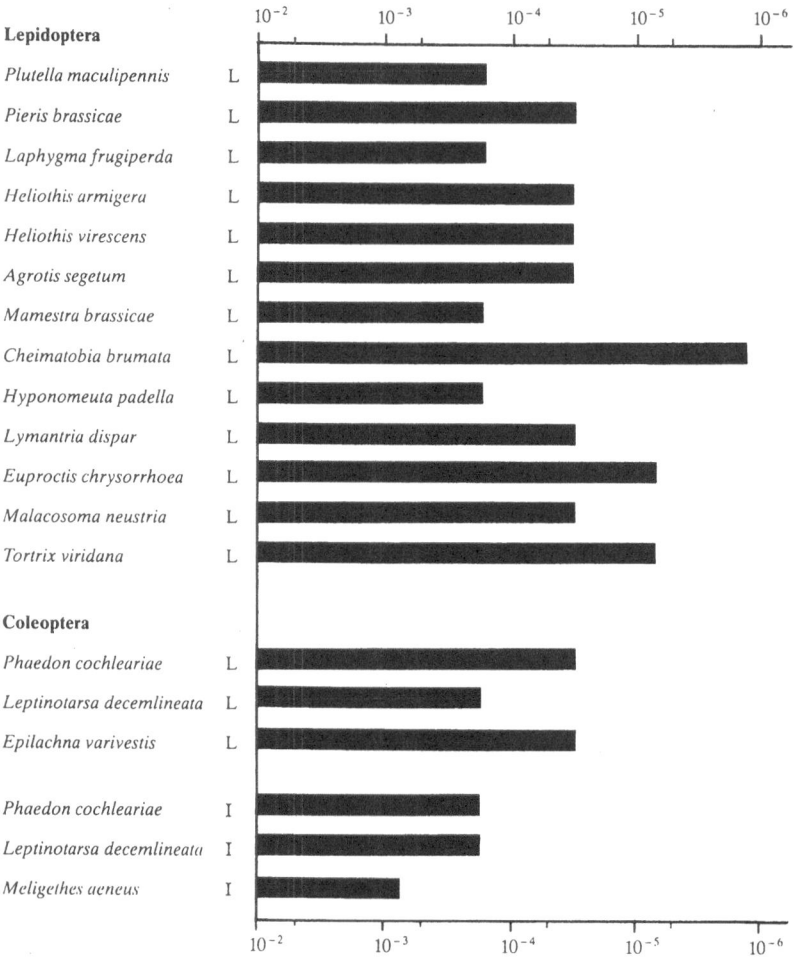

Fig. 14

aspects of pharmacodynamics of pyrethroids and isomers see the literature [234]. According to pharmacodynamic analysis the 1Rtrans isomer of bromethrin is intrinsically more active than the 1Rcis-isomer, taking into account metabolism and distribution [212], in spite of the higher toxicity of the cis-isomer in terms of LD_{50}.

For a review on the metabolism of pyrethroids in insects (69 references) see Soderlund [228] and [229, 230].

In addition to the afore-mentioned factors influencing the insecticidal activity of a pyrethroid, this activity varies, sometimes considerably, from species to species, or even within races, as found in investigations of 14 different silkworm races [211a]. Reasons for this may be found in different detoxications, access to the nervous system and binding characteristics at the target site in the nerve membrane.

A standardized leaf-dip bioassay at constant temperatures and light conditions, lasting for several days, shows this variation of toxicity of cyfluthrin for several species in Fig. 14.

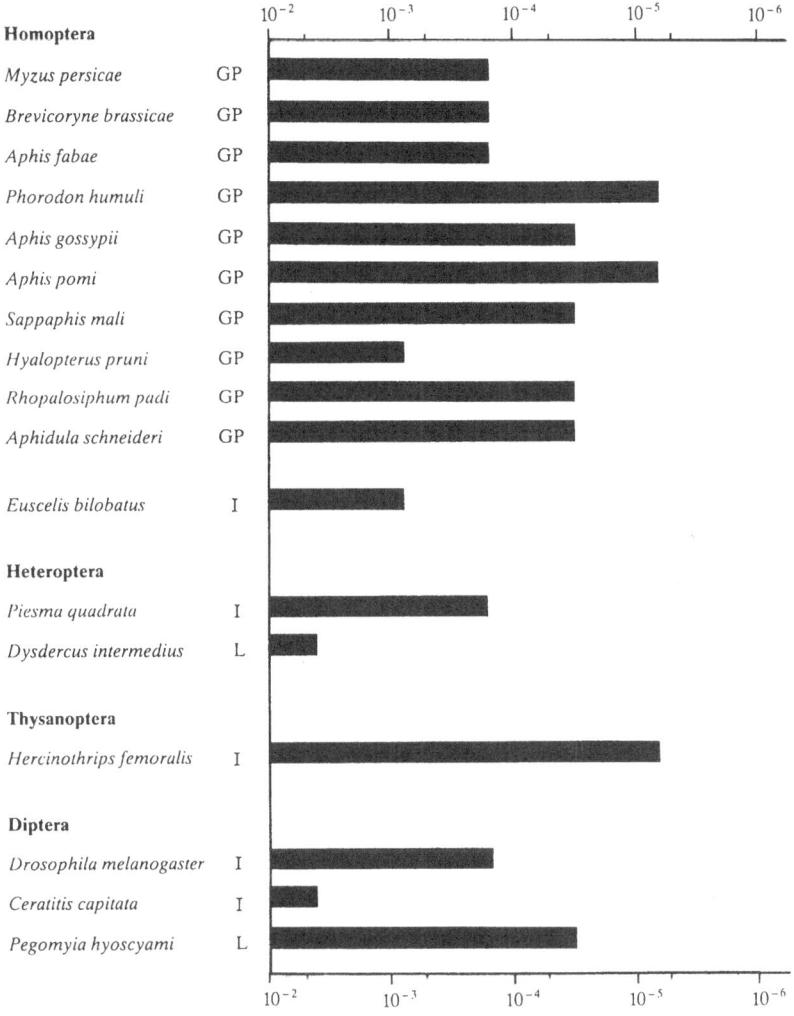

Fig. 14. Acute Susceptibility of different insect species to cyfluthrin (LC_{100}) [58]. x-axis: % active ingredient, l = larvae, I = imagines, MP = mixed population

Under field conditions however, this intrinsic difference of susceptibility may be even more pronounced due to the different way of life and feeding habits of insect species.

Another important part of pyrethroid action is the knock-down-effect, rapid in onset and sometimes very long lasting, but not always directly lethal. Lower doses lead to recovery, as shown in the protocol on the action of cyfluthrin on L_3-*Plutella* larvae [226] (Table 71).

The very fast action (Table 72, 73) against dipterans of the fastest acting pyrethroid so far, fenfluthrin [57, 92, 235], does not only depend on its volatility alone. The electrophysiological efficacy is also remarkable [236, 237].

Table 71. Protocol of action of cyfluthrin-knock-down, mortality and recovery of L_3-*Plutella maculipennis*-larvae after foliar application

% a.i.	10' % k.d.	dead	20' % k.d.	dead	40' % k.d.	dead	60' % k.d.	dead	80' % k.d.	dead
0.1	98		100		6	94	4	96		100
0.02	94		100		20	80	12	88	2	98
0.004	96		100		100		88		12	20	80
0.0008	86		100		100		100		100	
0.00016	48		86		100		100		92	
0.000032	2		6		16		30		80	
0.0000064							4		6	

% a.i.	100' % k.d.	dead	120' % k.d.	dead	140' % k.d.	dead	24ʰ % k.d.	dead
0.1	100							
0.02								
0.004	16	84	16	84	12	88		100
0.0008	90	10	90	10	90	10		100
0.00016	96		90		90		6	94
0.000032	90		80		80		0	14
0.0000064	6		6		6		0	2

Table 72. Diptericidal properties of fenfluthrin

(−)1R, trans

Biological effect		Animal	Min. effective dose	Mode of application
Toxicity LC$_{100}$		*Stomoxis*	3 ppm	Spray
		Lucilia	3 ppm	Spray
		Aedes	0.01 ppm	Spray
Inhalation toxicity				
speed of action	KD$_{100}$	*Aedes*	0.0025 mg/m^3	Air after 4 min
Lethal effect	LC$_{95}$	*Aedes*	0.00025 mg/m^3	Air after 24 h
	LC$_{100}$	*Aedes*	0.0025 mg/m^3	Air after 24 h
Speed of action	KD$_{95}$	*Musca d.*	0.025 mg/m^3	Air after 24 min
	KD$_{100}$	*Musca d.*	0.1 mg/m^3	Air after 20 min
Lethal effect	LC$_{95}$		0.05 mg/m^3	Air after 2 h

Table 73. Knock-down-effect KT (mortality M) of fenfluthrin-aerosol against sensitive and resistant dipterans

Application rate %	Aedes sens. min.		%	Musca sens. min.		%	OP-resistant min.		%	Pyrethroid res. min.		%
	KT_{50}	KT_{95}	M	KT_{50}	KT_{95}	M	KT_{50}	KT_{95}	M	KT_{50}	KT_{95}	M
0.0025	3'	7'	100									
0.005	2'	4'	100									
0.025				16'	24'	56						
0.05				10'	14'	93						
0.1				7'	10'	100	11'					
0.25							11'	19'	58			
2.5										21'		
5.0										13'	35'	14
Dichlorfos 0.25	25'	31'	100	24'	33'	99						

Table 73a. Mosquito vector controll with Cyfluthrin

	LC_{50}	
	Larvae mg/l	Adults (mg/cm^2)
Culex qu.	0.0007	4.5
Anopheles st.	0.0010	5
Aedes ae.	0.0230	18

The diptericidal tetramethrin is less effective against mosquitoes than against flies. *Culex pipiens pallens* is less susceptible to pyrethroids than *Aedes aegypti*, the vector of the yellow fever virus.

The important mosquito vectors *Culex q., Anopheles st.* and *Aedes ae.* differ in their high sensitivity to pyrethroids as shown in one Indian investigation with Cyfluthrin [237a] Table 73a.

Doses insufficient for lethal action are of practical importance, for instance in vector-control-programs. In view of the high activity of pyrethroids it is a very difficult technical problem to distribute the minute amounts of a few grams (or even less than 1 g [238]) evenly on the irregular surface on one hectare of a natural setting of an African river-bank. The dose applied as LD_{90} to controll mosquitoes was for example 0.3–5 g/ha. If one uses larger quantities of an aqueous carrier-medium, quite a loss of this small amount of active ingredients might occur due to adsorption on the larger surfaces of spraying equipment. On the other hand, the use of ultralow volume (ULV) poses the difficulty of uniform distribution of the spraying swath, this consist of very small droplets containing higher concentrations of the pyrethroid. With this method, a local underdosing may easily occur. In tests with tsetse flies with application of doses lethal to less than 50%, one observes a long lasting knock-down in the survivers as shown in the next table 74:

Table 74. Recovery time (h) for 50% of the surviving tsetse flies after topical application of a $LD_{15\%48h}$

Pyrethroid	mobility		flight	
	♀	♂	♀	♂
deltamethrin	20,6	25,5	30	**
bioresmethrin	3,6	2	6	3,5
permethrin	*	*	4	4
tetramethrin	< 1	1	< 1	2,3

*) not lost; **) not regained

In tse-tse flies, knock down is accompanied with a much increased obortion rate of eggs and larvae due to a general relaxation of maternal muscles [651].

This knock-down was followed by recovery to a large extent. The surviving flies, however are quite unfit to survive a hot African day. They eventually die also. This conditional weakness during or after extended knock-down is not directly linked with the neurotoxicity. Lack of moisture for instance or the vacuum cleaner of the housewife may be the cause for the actual death of the house-fly.

In gypsy moth larvae, surviving a cypermethrin treatment, a strong reduction or even disappearance of certain hemolymphal proteins was detected, the level of which was correlated with loss in viability and fecundity [239a].

For household purposes a very rapid knock-down-action against fly and mosquitoes is the desired property. Under optimal conditions there is a good chance for a knocked-down fly kept in a moist environment for a complete recovery. They first regain the normal posture, later the capability of jumping, while flight is resumed still later [240a]. In fact, the neurophysiological symptoms are longer lasting than the actual knock-down. The fly can live and fly with strong electrophysiological symptoms of a pyrethroid intoxication, like repetitive firing [240].

The knock-down-effect also depends on temperature. At higher temperatures, knock-down and recovery are faster, while at lower temperatures a higher mortality [241] is observed. For permethrin and allethrin this negative temperature coefficient is caused by pharmacodynamics plus target site interactions, but for cypermethrin only pharmacodynamics seem to be important. Knock-down-resistance does not seem to be dependent on temperature.

Side-effects of Pyrethroids on Pests and Non-target Arthropods

The action of pyrethroids against insects or mites is not restricted to the lethal activity and knock-down properties. Sublethal doses bring about several effects, seen in the laboratory and in the field. The effect of fenpropathrin on mites over a broad concentration-range is demonstrated in Fig. 15 [242]:
After pyrethroid treatment of a plant, the concentration of the active ingredient shows

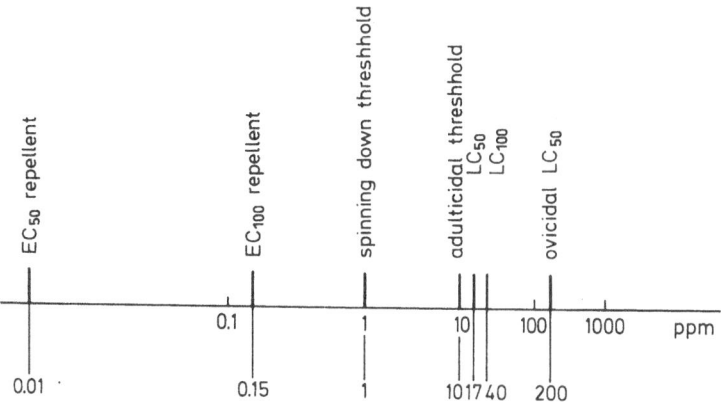

Fig. 15. Biological effects of fenpropathrin against mites over a broad concentration range

a sharp decline after a few days due to degradation, but an only slowly decomposing amount of pyrethroid remains on the leaves longer [242]. Sublethal concentrations are still effective as a feeding deterrent for larvae of *Pieris brassicae* [243] and *Drosophila* [244]. Repellent activity is often seen in only weakly toxic pyrethroids [245], like simple pyridylmethyl-chrysanthemates, or in completely inactive ones, like 1Strans permethrin [246] against black-fly, which attack cattle or in the R,R-fenvalerate-isomer [101]. Racemic permethrin and fenvalerate were recommended as effective mosquito-, fly- and tick-repellents [248]. However, this effect against most blood sucking arthropods was actually weaker than that of the standard repellent deet [248a]. In using permethrin as a repellent against cattle-horn-flies, it was even possible to treat only one cow to repel the black-fly from a whole herd [249]. However, in the meantime, there is now a widespread behaviour based resistance of the cattle-horn-fly [250], they are no longer repelled in large areas of the American south.

Vaporthrin may be usefull as a cockroach repellent [250a].

Another effect of this irritancy is the interruption of sex-signalling seen in *Pectinophora* [251]. On the other hand, the simple ethylchrysanthemate rhinolure is a sexual attractant for the coconut pest, the beatle *Oryctes rhinoceros* [251a].

Besides the repellent effect of permethrin, cypermethrin and others, sublethal doses negatively affect the growth, development and fecundity of *Spodoptera* [252] and aphids [253]. Moreover, ovicidal effects are observed in leafminers [254], *Plutella xylostella* [255], bollworm [256], *Spodoptera* [257] and others, depending on the age of eggs. Eggs of resistant insects are less susceptible [256]. In addition, the sublethal irritancy also inhibits oviposition [257].

In most cases, pyrethroids, are not sufficiently effective against soil-inhabitating-insects, apparently due to the low vapour-pressure and strong absorption by soil particles.One positive aspect of this behavior is, that pyrethroids, due to their low mobility in the soil are not prone to leaching into the groundwater, a negative side-effect of a number of classical pesticides, which is getting more and more attention. Highly hydrophobic pyrethroids, very rich in fluorine content, but lacking the CN-group like the pentafluorobenzylester fenfluthrin [258], the cis configurated tetrafluoro-4-methylbenzylester tefluthrin [259, 260, 260a] and bifenthrin [261], however, are about ten times more active against corn rootworm larvae than any other existing soil-insecticide, showing no adverse effects against earthworms. The analogous tetrafluoro-4-pyridylmethylesters are also claimed to be soil-insecticides [262]. On the other hand, *Agriotes* larvae, living in soil, but feeding on plant stems above the ground, are extremely sensitive to pyrethroids like cyfluthrin [226]. The predator of aphids, the earwig, is only partly affected by normal field rates [263], showing knock-down, but only limited lethal effects. Effects on other soil-inhabitating-animals such as root nematodes are claimed[264], but this weak action, if it exists at all, does not seem to be suitable for commercialization in comparison to existing nematocidal standards.

Beneficial species like the earthworm did not show adverse effects after treatment with cypermethrin or alphamethrin in amounts of 100 mg a.i./kg soil [265, 266].

3.2 Effect of Pyrethroids on Acarina-Pests

In the beginning of screening of synthetic pyrethroids against important agricultural and veterinary pests it became evident, that pyrethroids were much less effective against acarina pests, although not inactive, than against insects. This supports the common experience, that in a great number of cases acarina pests (mites, ticks) are distinct from insect in toxicopharmacological terms as shown by the great number of selective acaricides.

Later on, however, pyrethroids were discovered, showing subtle, but distinct structural differences in certain parts of the molecule, most of them containing fluorine, which proved to be outstanding acaricides. In addition to claims of poor insecticidal performance of some highly active pyrethroids, bad news came from field experimenters and fruit-growers about effects quite opposite to those desired, namely cases of the resurgence of mite-populations [267, 268, 269] after pyrethroid treatment. It even became common practice among experienced acaricide field testers in the case of a mite shortage in orchards to start the acaricide field test with a preceding pyrethroid treatment, in order to raise the spider mite population for a reasonably infested test plot. Moreover, a 10 fold increase of deposition of spider mite wintereggs was observed after application of e.g. deltamethrin or cyfluthrin [267].

These surges in mite population were attributed to the established stimulation of respiration rate [270, 271] (Table 75), shortening of the developmental

Table 75. Effects of different commercial pyrethroids on two-spotted spider mites

Pyrethroid	Toxicity	Effect on respiration of mites %	
	LC_{50} (ppm)	At beginning	After 3 days
Bifenate	50	+20	−70
Fluvalinate	100	+14	−46
Cypermethrin	130	+ 4	−46
Fenvalerate	160	− 8	−40
Fenpropathrin	240	+10	−65
Deltamethrin	290	0	0
Permethrin	320	− 8	−20
Flucythrinate	390	+14	−46

period by a factor of two and increase in fecundity [269]. Warm weather favours this effect very much [271].

There is a remarkably different ranking of the pyrethroids in Table 75 in terms of toxicity, and the effect on respiration rate. Pyrethroid-induced resurgence is also a problem in combatting the brown plant hopper, an important sucking insect pest in rice fields [272]. This is the main reason, why pyrethroids, at present, are officially not regarded as useful for this important crop.

However, there are pyrethroids claimed not to cause resurgence, e.g. fluvalinate [270a], which are supposed to retain good activity in addition to strong repellent properties [273]. The latter fact is another important facet of the effects of pyrethroids on mites, not necessarily connected with the intrinsic toxicity (Table 76).

Table 76. Ranking of acute toxicity and repellent action of pyrethroids toward *Tetranychus urticae*

Pyrethroid	Rank		
	Acute toxicity	Repellency	Overall activity
Deltamethrin	1	1	6
Cypermethrin	2	4	4
Flucythrinate	3	5	3
Fenpropathrin	4	3	2
Fenvalerate	5	6	5
Fluvalinate	6	2	1

The visible and commercially appreciated overall activity consists of two components: actual toxicity and repellency. For this two effects there are different rank-orders, and only a combination of both leads to the true rank-order and to the surprising result, that the least toxic compound in laboratory tests was the best one to protect a leaf from mite damage in the orchard [273].

The following pyrethroids are supposed to be especially effective on mite pests:

(S¦R'S')

The next table 77 shows some acaricidal data in connection with structural changes of pyrethroids:

Table 77. Structure-acaricidal activity data of some pyrethroids

X	Y	Z	W	Stereochemistry	LD_{50} Toxicity to Tetranychus (mite)	LC_{100} Boophilus (tick)	Lit.
	H	CN	F	1RcisZ, αS	5,4 ppm	0,1 ppm	[276]
	F	CN	F	1RcisZ, αS	more active		[277]
					relative		
F	F	CN	H	1Rcis, αS	100		[278]
Cl	Cl	CN	H	1Rcis, αS	1		
Br	Br	CN	H	1Rcis, αS	1		
Fenvalerate					1		
F	Cl	CN	H	(±)cis, α (RS)	2	6	[279]
F	Cl	CN	H	(±)trans, α (RS)	4	1	
Cl	Cl	CN	H	(±)cis, α (RS)	1	1	
F	F	CH₃	H	1Rcis, αR	10 ppm	6,2 ppm	[280]

Pyrethroids are usually not very active against cattle ticks. However, some α-cyano-compounds do have more or less higher activity than the organo-phosphates or amidines, the former standard compounds for this purpose. These pyrethroids are flumethrin, cyhalothrin, cypothrin, fenvalerate, cypermethrin and deltamethrin. The most active commercial pyrethroid recommended to be applied at 30–40 ppm against all kinds of cattle ticks and their larvae (0.01–1 ppm) is flumethrin [281, 282], useful for treatment in dips, spray and pour-on, where a single dose of 400 mg per animal is sufficient. Certain ticks are affected even with less than 3 ppm. This pour on method allows the protection of a whole herd of cattle, sheep, goats, etc, by treating just one or a few animals [282a]. In combatting ticks, long-lasting protection of cattle is desired. The next table 78 lists some of the required doses against ticks and against other arthropods (Table 79).

Table 78. In vitro inhibition of egg production of fully-engorged female one-host ticks by flumethrin LC_{99} (ppm) [283]

Boophilus microplus			
	Yeerongpilly	(OP-sens)	2,5
	Biarra	(OP-res.)	5
	Malchi	(DDT-res.)	9
	Ulam	(Amidin-res.)	6
	Porto Allegre	(OP-sens)	2
Boophilus decoloratus			
	Onderstepoort	(OP-sens)	5
Boophilus annulatus			
	Mexico	(OP-sens)	2,2

Recommended use rates:

One host ticks	30 ppm
Multi host ticks	40 ppm
Biting and sucking lice	30 ppm
Keds	30 ppm
Psoroptic mange	30 ppm
Safety margin (cattle, sheep)	> 50 × rec. use rates
Withdrawal period (cattle)	0 days, for meat and milk.

Table 79. The selectivity of (±) trans Z.E-flumethrin as tickicide in laboratory tests in comparison to the effects against other arthropods: [284]

Arthropod	approx-Toxicity LC_{90} ppm
Boophilus microplus	1
Tetranychus urticae	100
Myzus persicae	20
Phaedon cochleariae	1
Drosophila melanogaster	100
Plutella maculipennis (larvae)	10
Laphygma frugiperda (larvae)	50
Calandria granariae	500
Blatta orientalis	> 1000
Aedes aegypti (adults)	> 1000
Aedes aegypti (larvae)	1000
Musca domestica (adults)	> 1000

Cyhalothrin provides good protection with 70 ppm [285] in plunge dipping. Resmethrin is supposedly effective against the household mite *Ornithonyssus bacoti* [286], scabies and harvest mites which are parasitic on humans [287]. The use of permethrin as repellent for the lone star ticks has been successfully tested [288].

3.3 Factors Lowering the Efficacy of Pyrethroids Against Insects Resistance

As soon as the modern pyrethroids were applied commercially in agriculture and public health, they had, to the surprise of many, to face more or less resistant insects in increasing numbers in certain areas, where they had never been in use before [289, 290, 291]. This resistance showed up as a diminished knock-down-rate, and was in the first instances related to prior use of DDT over many years. On the other hand, the use of pyrethroids against Anopheles for example, rendered this pest 1000 fold more tolerant to DDT [292].

This cross-resistance does not follow a clear structure activity line. For example in one experiment [293] house-flies selected with DDT developed cross-resistance-factors to pyrethroids as follows:

bioresmethrin	53	
permethrin	11	
deltamethrin	115	
cypermethrin	41	αCN-Pyrethroids
fenvalerate	72	
fenpropanate	140	

α-cyano-pyrethroids tend to show a higher degree of cross-resistance, although not correlated to their activity against flies. The so-called kdr-(knock-down-resistance)-flies were found mostly in northern European isolated farms, where flies, spending the whole life-cycle in the warm environment of stables, suffer from intensive selection pressure by frequent DDT-treatment, while in warmer climates they also live outside and mix with the majority of susceptible ones. Another political and economically important selection of resistant mosquitoes came about from the longer lasting malaria eradication programs with DDT [294] in subtropical and tropical areas, in connection with the widespread application of DDT in ricefields. Here, the degree of DDT-resistance and resurgence of malaria was positively correlated with the increase in crop yield, leading to self-sufficiency in a basic crop in some important larger developing countries [295].

The DDT-cross-resistance of some insect species in the fields embraces all pyrethroids in use with different factors from 3 to 20 for example for mosquitoes [296]. In other species this resistance is more specific and depends on the structure of the pyrethroid.

In the Australian Emerald Valley 50% of the population of the main cotton-pest *Heliothis armigera* became resistant to the pyrethroids within one season, almost without any early warnings for most of the farmers, thus threatening the growth of cotton in that region. Resistance factors for permethrin, fenvalerate, deltamethrin and cypermethrin were 15–30 [297, 298]. The main Egyptian cottonpest, *Spodoptera litura*, gradually developed a 14 fold tolerance within 15 generations [299] in one experiment.

In another case the storage beetle *Sitophilus oryzae*, selected in the laboratory with permethrin or deltamethrin developed cross-resistance to other pyrethroids, which was linked only to one sex [300], and did not affect the onset of knock-down, rather the degree of mortality.

Widespread failure to combat the parasitizing cattle-horn-fly with pyrethroid-containing ear tags were reported after initial very good results with fenvalerate and permethrin in 10 states of the United States. Here, insensitivity-factors between 34 and 48 were observed [301, 302]. This resistance, attributed to the kdr mechanism caused by nerve insensitivity, was only incompletely recessive, and was determined by a single, sex [303] linked gene [303a].

In 1985, the pyrethroid sensisitivity of furniture carpet-beetle, in different laboratories and living on wool-fabrics, varied by the factor of 4. The common moth did not show this variation [304].

Over a period of 9 years, the cases of observed pyrethroid resistances under practical conditions increased from 6 species in 1976 to 70 in 1985 [289, 290, 291]. These species included pests and beneficials. Complete control-failure by pyrethroid, although only of local economic importance, has been reported for *Spodoptera* in Turkey, *Plutella* (Far East), *Heliothis* (Australia) and fleas, flies and lice in many places [305]. 1987 there were two closely watched spots of pyrethroid resistance in the cotton area of the Rio Grande Valley (R-factor > 100 and 5–30) and one in the Bravos Valley of Texas. (R-factor 250). The outbreak of resistance in columbian cotton areas also deserves carefull attention. Particularly, Asian *Plutella* (Thailand, Japan) [306] in the cabbage fields developed a strong multi-resistance against pyrethroids, organophosphates and carbamates, leaving this important crop virtually unprotected, thus leading to a loss of investment and earnings for the farmers.

Whether the *Thrips tabaci* and *Thrips nigropilosus* occurring as natural pests on chrysanthemum flowers in pyrethrum plantations are tolerant to the natural insecticide remains to be shown [38].

Understanding the conditions which favour the development, the causes, and the mechanism of resistance are a crucial challenge for the future of pyrethroids.

For fly-control, close investigations revealed, that pyrethroids with long residual activity in indoor application in cooler climates speed up the development of resistance due to constant selection pressure [307]. The not so frequent use of faster acting, more volatile and easier disappearing pyrethroids as space sprays is therefore recommended for management of a resistance-potential in housefly and mosquito [308, 309, 310]. Short, but effective periods of action of a relatively fast disappearing pyrethroid kills only adult flies, leaving the emerging larvae without selection pressure and may give immigrating susceptibles flies a chance to restore a susceptible population. This strategic idea became official fly-resistance-management-policy in Denmark, leading to a restriction in the use of residual insecticides on farms. By this method completely uncontrollable fly populations could be controlled again [309] with pyrethroids.

Compounds, useful for this approach, are fast acting pyrethroids like pyrethrum, allethrin, bioresmethrin and prallethrin [311] and similar compounds. A recent structure-resistance-factor analysis revealed, that metaphenoxybenzyl esters favour the occurrence of super-kdr in housefly. Benzylfurylmethyl esters, and far less so esters of cyclopentenolons and pentafluorobenzyl alcohol, do not favour this strong type of resistance [310a, 335].

A prophylactic policy is followed in Egypt, where a strict rotation of different insecticides following an outlined official schedule is supervised by the government. Similar actions ("window strategy") have been taken on a more voluntary basis in Australia.

A constant monitoring of resistance levels is necessary. For this reason simple field tests are under development, e.g. [335a].

As the underlying mechanism for developing pyrethroid resistance several factors have been found.

a) The knock-down-resistance (kdr), or more so, the super-kdr, was found to be caused by *nerve insensitivity* [312, 313, 314] due to a change in the actual binding site for the pyrethroid in the nerve-membrane. For the neurophysiological aspects see Sect. 7.1.3.

In European flies, it seems to be the predominant resistance-mechanism, but has also been found in mosquito larvae, *Anopheles,* different *Heliothis-* [314a], *Plutella-* and *Spodoptera* [315] species. Kdr is programmed by a recessive gene not responsible for metabolism [316, 317]. This gene is localized on chromosome III of *Aedes aegypti* and is allelic to one of several DDT-resistance-factors [318] yielding a 20 fold resistance to permethrin [319]. Lower resistance was found in most cases with pyrethroid esters bearing the 3-phenoxybenzyl-moiety as the alcohol component. However, there is no clear structural relationship to resistance. Nerve insensitivity due to an architecturally or functionally altered target site is proposed for resistance to permethrin or transpermethrin in mosquitoes [320, 321] and German cockroaches [322], being resistant to allethrin, permethrin and fenvalerate, but not to deltamethrin. This was connected to the type I-Pyrethroid mode of action. Table 80 lists data of resistance [316] in heterozygous and homozygous mosquito larvae against several highly active pyrethroids.

Target site insensibility is related to the nature of phospholipides in the nerve membrane, which alter the conformation of membrane bound proteins, presumably sodium channel proteins, without disturbing the physiologically necessary function of this ion channel. Super-kdr-flies are very healthy. Susceptible flies have liquid membranes at 20 °C, whereas the membranes of super-kdr-flies are quasi crystalline at that temperature. Nerve membranes of susceptible flies show a phase-transition-temperature of 14 °C, kdr-flies at 19 °C and super-kdr-ones of 21 °C [323].

b) While increased *oxidative metabolism* in certain houseflies [324] or hornflies [325] is only a small part of the observed resistance, other resistant fly-populations in areas of kdr-resistance developed additional oxidative and hydrolytic deactivation [314]. Particularly the mechanism due to increased mixed functional oxidase activity(MFO) became more and more important as shown in many recent cases. In general,with increasing resistant factors one has to expect that, in most cases there is a multiple genetic cause for resistance, each varying in importance; but oxidative metabolism will play a major role in cases of very high resistance.

A complex cause was found in the up to a 100 fold pyrethroid resistant *Heliothis armigera* from Thailand [326], a reason for cutting down the cotton acreage in this country between 1980–86 to 1/3 [327]. *Plutella* larvae from Thailand failed to respond to pyrethroids because of increased oxidation, since the MFO-inhibitor piperonylbutoxide (PBO) lowered the resistance, but not so the esterase inhibitors [328]. In this case the organophosphate methomyl even showed a negative cross-resistance, being more active against the pests, resistant to permethrin, cypermy-thrinand deltamethrin, but not to fenvalerate. However, after a further selection with pyrethroid even PBO and methomyl together lost efficacy [329, 330].

Table 80. Resistance factors of $\pm$trans permethrin resistant mosquito larvae to very active pyrethroids

Compound		LC_{50}(ppm) susceptible strain	Rf LC_{50} homozygous strain	Rf' LC_{50} heterozygous strain
($\pm$)cis/trans RS	Cyfluthrin	0.30	253	6
1Rcis	1R cis Permethrin	0.36	750	
1Rtrans	Fenfluthrin	0.55	63	6.5
1Rcis S	Tralomethrin	0.58	1190	13
($\pm$)cisZ, RS	Cyhalothrin	0.73	301	3.5
($\pm$)cis, Z	Bifenthrin	2.1	19	7
($\pm$)trans Z		2.3	140	17

Table 80 (continued)

Compound		LC$_{50}$(ppm) susceptible strain	Rf LC$_{50}$ homozygous strain	Rf' LC$_{50}$ heterozygous strain
(±)cis		2.8	19	7
1Rtrans	1R trans Permethrin	3.0	1200	7
(±)cis		3.5	28	10
(±)cis		4.8	13	3
S; RS	Flucythrinate	5.6	300	7
(±)cis		5.6	1600	14
		5.7	130	7.5

Similarly permethrin resistance in the housefly due to oxidative metabolism was found in cases in Japan [331], Canada [332] and the USA [333], where the resistant strain had developed a much higher degree of resistance to the trans-isomer of resmethrin than to the noncommercial cismethrin due to oxidative inactivation. The usually greater importance of oxidative metabolism of pyrethroids in insects is supported by the findings, that the non-hydrolyzable ether type of pyrethroids, such as a variation of etofenprox with increased activity,

is unable to brake multiple resistance against organophosphates, pyrethroids and super-kdr in houseflies [334].

The highest resistance so far has been found in the American housefly strain Learn PyR which showed an hitherto unseen level of resistance of more than several thousand fold to 3-phenoxybenzylester-pyrethroids, as listed in the next Table 81 [335]:

Table 81. Resistance factors of 3-phenoxybenzyl pyrethroids as found in the resistant housefly strain Learn PyR

Compound	LC_{50} activity against susceptible flies	Level of resistance
Deltamethrin	0.002	> 10000
Cypermethrin	0.007	5100
Cyhalothrin	0.007	4700
RU 38702	0.013	9200
Bioresmethrin	0.022	2000
Permethrin	0.024	6200
Bifenate	0.027	560
Cyfluthrin	0.042	240
Fenfluthrin	0.048	42
FMC 54617	0.054	2000
Phenothrin	0.058	> 3400
Fenvalerate	0.084	> 2400
Flucythrinate	0.096	930
Fluvalinate	0.19	> 1000
	0.63	170

The introduction of the 4-fluorine substituent lowered this resistance 20 fold, and other alcohol components like pentafluorobenzyl or allethrolon in the pyrethroid did not show this high degree of resistance.

The major cause for this high level resistance was found in a metabolizing gene on chromosome I, which causes an increased abundance of the whole oxidizing machinery, comprising cytochrome P 450, cytochrome B_5 and its recharger NADPH-cytochrome-P-450-reductase. In addition to this increased MFO-activity, target site insensitivity and reduced penetration are also involved. Genetic analysis of this strain by separation in different homozygous strains revealed the origin of the other causes and their relative contribution to the overall resistance, as ranked in the following listing [336]:

Gen	factors of resistance
5	increased cytochrome P 450 increased NADPH-Cytochrome-C-reductase
3	kdr reduced penetration increased NADPH-cytochrome-C-reductase
1	increased cytochrome P 450 increased cytochrome B_5
2	increased cytochrome P 450 increased cytochrome B_5

A wild strain of pyrethroid resistant *Heliothis virescens* showed not only a higher degree of metabolic activity, without producing novel metabolites, but also reduced neuromuscular sensitivity in electrophysiological experiments [337].

Epoxidation of chrysanthemic esters at the acid-moiety may yield 2-isomeric epoxides of equal or much reduced neuroactivity and toxicity, respectively. The isomeric 1-R-trans-tetramethrin epoxides differ in their neuroactivity at the cercal sensory receptors of American cockroaches by a factor of 10^5 [338].

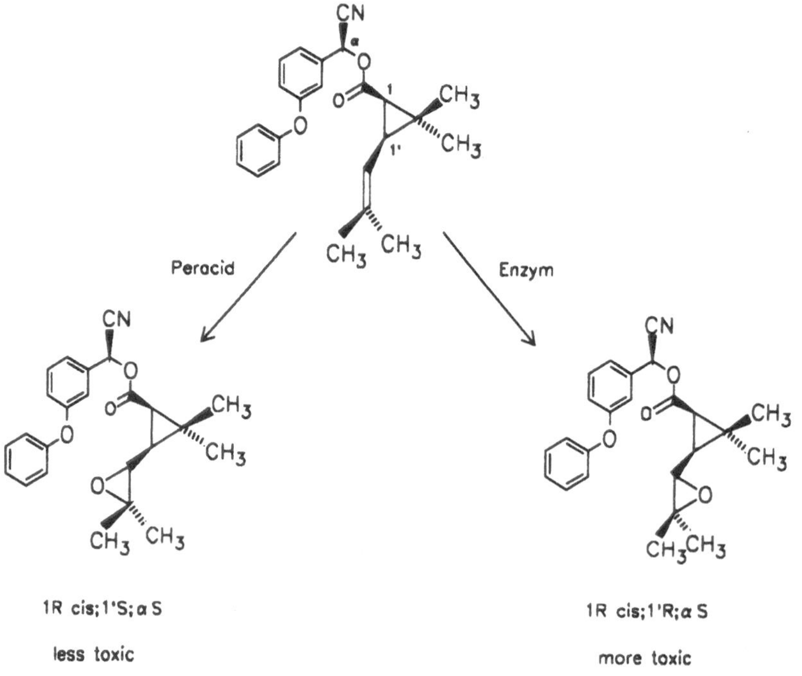

Fig. 16. Different sterochemical course in the oxidation of cyphenothrin by peracid or enzymatic oxidation

In the case of 1RcisαS-cyphenothrin the enzymatically formed 1'-R-isomer of the epoxide as shown in Figure 16 is 30 times more toxic to mice than the 1'-S-diastereomer, formed predominantly by in vitro standard peracid oxidation. On the other hand, the last one causes stronger repetitive firing in nerve preparations than phenothrin [74]. The reported absolute configuration of the active epoxide is interesting with regard to the proposed active conformation of pyrethroids (p. 79). There are hints, that the metabolism of phenothrin in mamals occurs without formation of these epoxides [74a]. If an insect is to be protected by an oxidase, which selectively produces only the inactive isomeric epoxide, certainly this would be a good means of metabolic protection against the poisonous action of a chrysanthemate pyrethroid.

Insects feeding on plants containing toxic alkaloids may have a higher disposition for developing pyrethroid resistance. Solanine in potato leaves induces the MFO in Colorado potato beetle [339]. Therefore the breeding of potatoes for a higher solanine content as 'natural insecticide' actually favours the development of resistance and leads to even diminished protection of this crop, thus increasing the pest problem. The induction of MFO in caterpillars by the natural food seems to be a common phenomenon [359, 645]. Phenobarbital, a strong inducer of MFO acts therefore as antisynergist [125].

By the same token, the LD_{50} for deltamethrin in larvae of *Spodoptera littoralis* was about two times higher after pretreatment with a non lethal dose of the insecticide lindane, which supposedly also induces a higher esterase level [835].

c) If *increased hydrolysis* is involved, then inhibitors of esterases act as synergists, as found in the case of the naturally occuring high resistance of the predator green lace wing (*Chrysopa carnea*), preying on *Heliothis*. It is almost one million times less sensitive than a housefly [340]. Here, the non-toxic P-ester phenylsaligenin blocked the highly active cis-pyrethroid esterase, resulting in a 70 fold increase in toxicity [340] to this beneficial insect. Egyptian *Spodoptera litoralis* larvae in cotton developed resistance due to 3 to 6.5 fold higher esterase activity [341].

The predator mite *Amblyseius fallacis* became resistant to permethrin due to higher esterase activity [342]. In the resistant Singapore *Plutella xylostella*, 1600 fold resistance to deltamethrin, and 30000 fold resistance to cypermethrin was observed. While deltamethrin in the susceptible strain was insufficiently synergized, synergism in the resistant strain was much higher [343]. Pyrethroid resistant *Porinia* moths rapidly metabolize the toxicant by esteratic and oxidative cleavage [344].

In one American case, houseflies became more resistant to trans-permethrin than to the cis-isomer. This resistance was only partly abolished by the oxidase-inhibitor PBO and not by the esterase inhibitor DEF [333].

In the resistant strains of larvae of cattle tick (Malchi and DDT-R-strain) two to three independent factors are operating: The first one unaffected by esterase-inhibitors, results in decreased knock-down by permethrin, the other one, apparently due to increased esterase activity, causes diminished kill of the ticks by permethrin, cypermethrin or deltamethrin in the Malchi-strain [345]. This was overcome by coumaphos synergism. Russian potato beetles were not sufficiently sensitive to permethrin due to elevated esterase-, gluthatione- and cytochrome P 450 levels [345a].

d) There is an interesting and different way of inactivation of a pyrethroid by certain insects. Not by metabolic degradation, but by sequestering it in certain stores, a toxic concentration at the nerve is avoided. This may be the cause of resistance in *Wiseana cervinata* larvae [346]. A resistant *Myzus persicae* produces a carboxy-esterase in excessive amounts, not, however, to hydrolyze the toxin, but to sequester it unchanged. Only the inactive 1-S-trans permethrin is rapidly hydrolyzed and released from these storages [347]. The binding of the toxic isomers to this non-target site could be comparable to the binding at the toxicologically relevant neural target site. May be, some of the 'unspecific' binding of pyrethroids in the organism is actually highly specific, but without acute physiological consequences.

e) Moreover, cypermethrin penetrates more slowly than permethrin in the more resistant Malchi strain. Reduced penetration is also one of the causes for resistance in one pyrethroid resistant housefly strain from France [348].

   The fact, that penetration may also contribute to the resistance complex was supported by an American housefly strain, in which trans-permethrin penetrated slower in the resistant strain than in the susceptible one [336]. The main burden of resistance in houseflies, however, rests with the insensitive neural target site as seen in most reported cases so far.

f) Hints to a behavioral change as cause of failures to control pests with pyrethroids came from observation with the cattle hornfly *Haematobia irritans* [349], when the flies became hypersensitive to the repellent activity of fenvalerate and permethrin in addition to the development of lethal resistance to fenvalerate, permethrin and deltamethrin [350]. Pyrethroid resistant houseflies rest more briefly on insecticide treated surface [351] than susceptible ones.

   The outbreak of mite infestations in orchards after pyrethroid treatment, as mentioned before, was also connected to a behavioral change due to repellency, hyperactivity, reduced feeding and emigration to less dangerous places [352].

For a more detailed review of pyrethroid resistance see Sawicki (1985, 205 references) [391]. For the neurophysiological aspects of pyrethroid resistance see Sect. 7.1.3. Interesting observations have been made of the occasional occurrence of negative cross-resistance, when insecticidal compounds are more active to pyrethroid resistant strains than to susceptible ones, as reported for the α,γ-unsaturated isobutylamids [353].

Pyrethroid resistant insects are far more sensitive to the Nuclear Polyhedrosis virus of Mamestra brassicae than the susceptible insects [358a].

In most cases these resistant insects are less viable than the former susceptible ones. Without selection pressure the susceptible ones eventually will again dominate the population within a certain time interval. However, pyrethroid treatment will within a shorter time restore resistance.

Besides reduced sensitivity of insects to pyrethroids (=resistance) there is the simultaneous action of other drugs in insects, which reduce the efficacy against insects in an antagonistic way. In this manner the non-insecticidal steroisomers of cyphenothrin and resmethrin behave as antisynergists [354, 19]. Insect toxicity-data of this kind are shown in the following Table 82. See also p. 167.

Table 82. Topical toxicity of resmethrin single isomers and different isomeric mixtures. Additivity and antagonism in the insecticidal effect against housefly [19]

Isomer	LC$_{50}$ topical fly µg/fly
1R trans	0.013
1S trans	1.77
1R cis	0.041
1S cis	4.0
±trans	0.024 Additive
±cis	0.95 Slight antagonism
1R cis + 1R trans 3:7	0.016 Additive
±cis/trans 3:7	0.05 Antagonism

In nerve cell preparations the insecticides bioresmethrin, cismethrin and cis-permethrin antagonized the action of deltamethrin or kadethrin on open Na-channels [355]. 1S trans-tetramethrin prevents the action of the corresponding 1R trans-isomers on the nerve membrane in a noncompetitive manner, while the 1S cis-isomer competitively inhibits the 1R cis-isomer [356] in its action. Antagonistic action was also observed with endosulfan and dimethoate in a mixture together with deltamethrin [357] in controlling African thrips.

The selective acaricide azocyclotin, an inhibitor of oxidative phosphorylation, abolishes the acaricidal activity of cyfluthrin [358].

In contrast to certain other insecticides with a different mode of action the pyrethroids fenvalerate and permethrin proved to be less toxic to mosquitoes as long as they are blood-fed [357a].

A more trivial, but economically important cause of limited activity against certain insects is the limited access of the insecticide to the insect. This may be caused by the physical properties of the surface to be treated, which may absorb or degrade the active ingredient. This applies also for the surface of a living organism like a tick, onto which different pyrethroids have a different tendency for deposition, as found for trans-cypermethrin and trans-flumethrin. This difference causes the latter to express only one third of its actual potency [839]. Another cause may be the way of life of some insects being hidden under the protecting plant cuticula, or in niches of flowers, leaves, bark or other parts of the plant. Here, the lack of systemicity of pyrethroids greatly hampers the use of pyrethroids against certain pests.

3.4 Factors Enhancing the Action of Pyrethroids Against Insects

Pyrethroids are more or less suitable substrates for hydrolyzing or oxidizing enzymes in insects, depending on the developmental stage [359a] and species of course. Therefore one can expect enhanced toxicity caused by inhibition of these enzymes e.g. by organophosphates or PBO. On the other hand,a potentiating effect due to other neurotoxic effectors in addition to the complex action of pyrethroids may be another principle of synergism.

3.4.1 Metabolism-inhibitors as Synergists

3.4.1.1 Esterase Inhibitors

It is frequently claimed, as in patents [360, 361, 362, 363], that *esterase inhibitors* like certain insecticidal organo phospates (particularly those, less insecticidal, but non-polar) [360a] act as synergists for some pyrethroids in having more than the additive activity of the single insecticides. This synergism however was not found in malathion-permethrin mixtures [364]. Apparently the inhibition of gut esterase only results in synergism, as found with profenofos and to a lesser extent azinfosmethyl or monocrotophos for the easily hydrolyzable trans-permethrin in *Spodoptera littoralis* larvae [365].

In the greenhouse, profenofos, monocrotofos and methidathion potentiated the toxicity of cypermethrin 20 fold against whitefly [363]. Action of trans-permethrin or cis-cypermethrin in *Musca* or cabbage looper larvae is not or only weakly synergised by topically applied profenofos, sulprofos or DEF. However, an ingested P-ester raises toxicity up to 20 fold in case of cis-cypermethrin, but not so much for trans-permethrin [366]. Again, several specific gut esterases are inhibited to different degrees. Insecticidal methylcarbamates are also claimed to potentiate the action of pyrethroids [367, 368, 369, 370, 371], even more so in the presence of an organo-phosphate [547a].

Despite the potentiating action of some P-esters to pyrethroids in the laboratory and closed systems, it is hard to find, with any degree of certainty this mutual potentiation in the field, which would save expensive pyrethroids. To guarantee protection of the crop, a slight overdose is necessary, so that subtle effects are not easily seen.

It depends very much on the insect species, whether hydrolysis or oxidation is the most important pathway for pyrethroid detoxification. Yet, most of the evidence so far shows, that oxidation is the most common metabolic pathway in insect.

3.4.1.2 Oxidase Inhibitors

More or less effective inhibitors of microsomal mixed functional oxidase (MFO) are insecticidally inactive methylendioxybenzene derivatives like piperonylbutoxide (PBO) and many more natural or synthetic compounds bearing this structural feature [372]. PBO is the widely used pyrethroid-synergist only for household and laboratory use. PBO has no influence on knock-down-rate of cypermethrin in the housefly [373]. With higher rates however, even in outdoor field tests in Maine, synergism against Colorado beetle in tomatoes could be found with permethrin or fenvalerate [374]. This is interesting, however, only for cases of low levels of resistance [375]. Here, 56 g/ha fenvalerate plus 110 g PBO were equally effective as 224 g/ha fenvalerate alone.

Another more efficient MFO-inhibitor with stronger synergistic properties towards pyrethroid are certain propargylphosphates [376] and longchained alkylimidazoles

[344]. Synergism is greatest in the 3-phenoxybenzylesters [344a] in the case of mosquitoes.

Inhibition of oxidative degradation instead of blocking the oxidating reagent can also be accomplished by the stabilization of the substrate by taking advantage of the isotopic effect on CH-bond-strength by H-D-exchange. The result thereof is shown in the next table, where deuteration at certain positions of the molecule caused an impressive rise in intrinsic toxicity for insects [66].

Table 83. Metabolic stabilization by deuteration in

Increased topical toxicity against three insect species of DDT-pyrethroid hybrid molecules caused by hydrogendeuterium exchange initiated metabolic stabilisation

R^1	X Y Z W R^2	LD_{50} mg/kg topical			Lit.
		Blowfly e (synergized)	Cockroach	Budworm	
EtO	H H H H H	1.7 (0.11)	6.2	4.6	[66]
EtO	D D H H H	0.9 (0.07)	3.6	0.8	
EtO	H H H H CN	0.8 (0.03)	3.3	0.6	
EtO	D D H H CN	0.4 (0.01)	16	0.08	
EtO	H H H H C≡CH	0.5 (0.04)	3.6	2.7	
EtO	D D H H C≡CH	0.3 (0.03)	2.2	1.3	
EtO	D D D H C≡CH	1.2 (0.04)	2.0	0.8	
EtO	D D H H C≡CD	0.8 (0.04)	4.7	0.8	
EtO	H H H F C≡CH	0.002	0.02	0.005	[377]
CH_3	H H H F CN	3.5 (0.04)	2.7	0.7	
CD_3	H H H F CN	1.4 (0.037)	2.4	0.9	
Cl	H H H F C≡CH	0.3 (0.03)	0.9	0.4	
Cl	H H H F C≡CD	0.2 (0.03)	2.1	0.2	

The strongest effect of deuteration was seen on the susceptibility of budworms. They have the highest endogenous MFO activity (strongly induced by their natural food). Also interesting is the 10 fold increase in insecticidal efficacy of etofenprox, a new ether-type pyrethroid, in the presence of copper-II-oxide-carbonate. It remains to be seen, whether this has some connection to oxidizing processes [95].

3.4.2 Other Synergistic Mixtures

Certain combination of insecticidal stereoisomers of pyrethroids [378, 379, 380, 381] or of active and inactive diastereomers [382] of allethrin and combination of bioallethrin and bioresmethrin [383], fenfluthrin and vaporthrin [384], fenfluthrin and allethrin[385] and other mixtures of pyrethroids are also examples of synergistic mixtures. However, certain other mixtures of isomers are antagonistic (p. 108).

Finally, the high synergism with the octopaminergic chlordimeform in *Heliothis vir.* [387, 742] or the synergistic joint action of growth regulators like the chitin synthesis inhibitor [387] diflubenzuron or the juvenile hormone mimic RO 135223 [388] may be worth mentioning. However, methopren, a very close mimic of the natural JH is not such a potentiator. Trioxabicylo octanes, noncompetitive antagonists at the chloride channel, are claimed to speed up the knock down rate of pyrethroids [388a]. On the other hand, the chloride channel effector lindane also a noncompetitive gabergic antagonist renders deltamethrin less toxic to *Spodoptera l.* larvae [835].

A common experience of cotton growers in Central America is the enhancement of pyrethroid action by toxaphene or certain mineral oil fractions, as experimentally confirmed for cyhalothrin [389]. Other plant and mineral-oils also render pyrethroids more toxic to insects [390a, 390b, 390c]. The reason is unknown as yet.

Another interesting observation is the strong synergism of deltamethrin and the phototoxic dyestuffs fuchsin and methylenblue, and of flucythrinate with xanthine or eosin [390].

3.4.3 Metabolic Activation

Slow toxification of the pure, almost inactive isomer S,αR-fenvalerate occurs in the cabbage armyworm and *Aedes* [101], but not in *Musca*. It is open, wether this metabolic activation is an likely isomerization at the α carbon (see Sect. 3.2.2 Vol. 5 of this series) or another process. The efficacy was also demonstrated in field tests.

3.4.4 Improvement of Performance of Pyrethroids by Physical Means

The natural pyrethrum and by the same token allethrin are sensitive to light and air, and elevated temperatures [392, 393, 394]. Under the influence of light the alcohol racemizes and the side chain cyclizes, rendering the molecule inactive [395, 396]. For stabilization [415], and to prevent the formation of unpleasant odors or coloured and therefore lightabsorbing byproducts [397], antioxidants [398–405], phenylsiloxanes [405a], sulfidoesters [405b], UV-absorbers [406], azodyestuffs [407], benzophenon [408], para or ortho nitrophenols [408a], salicylic acid [408b]special dyestuffs such as organic cations e.g. methyl green on a clay carrier [409, 409a] such as montmorillonite, desactivated anilines [410, 411, 412] and general lightabsorbers for the wave-lengths between 300–400 mμ [413–414] are added. Synergists have an added task as stabilizers for pyrethrum [416, 414]. Other admixtures like boron compounds [417, 418], silicagel [419] and other inorganic carriers [420] serve the same purpose. A peculiar combination in connection with increased activity as a stomach poison on one side and diminished contact toxicity to beneficial insects on the other is accomplished by a 1:2 clarthrate of allethrin [421] and other pyrethroids [422] with cyclodextrin. The volatility of allethrin is lowered by the addition of higher boiling glycerol esters, polysiloxanes [422a], alcohols, squalene and the like [423], thus extending the residual activity. Microencapsulation, optionally with light-absorbing polymers [414a], of pyrethrum, permethrin or resmethrin certainly is an interesting proposal [424, 425]. Occasionally, metal-corroding properties of pyrethrum are mentioned, leading to the loss of the active ingredient in the metal container. Mercaptobenzthiazol inhibits this effect [426].

The overall activity of a pyrethroid under practical conditions of usage can be greatly influenced by the method of application. In the laboratory the usual formulation is a concentrate of the very hydrophobic active ingredient in an organic solvent, together with certain surfactants, rendering the insecticide emulsifiable in water. In general, this is more or less also the situation found during common agricultural practice on a farm, when a ready-made commercial concentrate is poured in a given amount of water for spraying on the crop. During application droplet size also plays an important role for desired and undesired effects. The application of the original concentrate of the pyrethroid without further dilution with water by ultra-low-volume-application (e.g. 0.5 l/ha) is an interesting new method, specially adapted to the characteristics of pyrethroids.

However, because of the sometimes less favorable properties of certain formulations of pyrethroids in some organic solvents which occasionally cause skin irritation for sensitive people, the development of water-based formulations for liquid or crystalline pyrethroids with no or almost no content of organic solvent attracted interest, not just for ecological reasons [427–430]. These aqueous formulations combined with an emulsifier are also of interest for indoor application as sprays [431], and essentially reduce skin irritation. This undesired side-effect can also be overcome by using vegetable oils [432] particularly tall oil, or special petrol-fractions like white oil [433] as solvents for pyrethroids [432].

For certain applications in the home or in animal husbandry a slow release of the active ingredient over an extended period is needed. Therefore, solid formulations of pyrethroids on the basis of organic polymers for use at ambient temperatures [434–437] as films or eartags were developed. But also inorganic solids like carbon black [438] or granular clay [439] are suitable for delivering pyrethroids for special purposes without loss of activity.

Combatting ectoparasites on animals needs special methods. Although pyrethroids like the tickicidal flumethrin are sufficiently stable under the conditions of a cattle dip, pour-on formulations [440–445] are much more efficient, where only a small amount of a specially prepared pyrethroid-containing mixture spreads out over the whole surface of the animal within a short time, thus giving complete protection. Recently it was found, that cypermethrin rapidly (10 cm/h) penetrates the skin of sheep radially, and that a lateral movement within the stratum corneum, together with the sweat/sebum emulsion may explain the dispersion of the active ingredient in the skin [449a].

An unusual proposal for tick control is the use of macroencapsulated preparations of cypermethrin for a feed-through treatment of tick infected cattle [450], giving a sufficient concentration of a.i. in the bloodstream of the cattle via passage through the gut walls. A blood sucking tick thus may receive its lethal dose with the blood meal.

The rapid action of pyrethroids is highly desirable for combatting indoor pests. Particularly the more volatile pyrethroids are suited for fumigant action. Besides spraying a diluted solution of a pyrethroid in a hydrocarbon or other volatile organic solvents using pressurized gas in the classical spray-can, thus providing a fine aerosol, technically simpler methods are of interest. Dimethyl ether as a propellant has an additional advantage in its rapid anesthetic and stunning effect, which allows the safe use of pyrethroids against stinging insect colonies, such as wasps [450].

An effective aerosol may be produced using a slowly burning solid mixture, which contains a certain amount of active ingredient which evaporates or is a part of the smoke being formed. Traditional methods using this system come from East Asia: foggers, mosquito coils and candles. Heat providing materials may be wood dust, carbon black [451], stearin [452], nitrocellulose paper [452a] pyrophorus iron [453], supported by smoke generating compounds [454] in connection with incenses like funori powder. This method also burns a part of the insecticide, but this can be reduced by addition of an antioxidant [455]. Still milder evaporation methods, which save the expensive active ingredient and avoid the pyrolytic development of other

Table 84. Physical properties of some pyrethroids [460, 461, 462]

Compound	$\lg P_{Oct H_2O}$	Calcul. solubility in H_2O (ppm)	Vapour pressure mm Hg/20 °C
MTI 800	9.5		
Flumethrin	8.6	0.0003	$1 \cdot 10^{-10}$
Phenothrin	7.4	0.01	$4 \cdot 10^{-6}$
Cypothrin	7.4	0.006	$3 \cdot 10^{-9}$
GH 380	7.0	0.02	$3 \cdot 10^{-10}$
Tralomethrin	6.9	0.01	$3 \cdot 10^{-11}$
Flucythrinat	6.8	0.01	$1 \cdot 10^{-9}$
Ethanoresmethrin	6.8	0.03	$3 \cdot 10^{-6}$
Permethrin	6.8	0.03	$3 \cdot 10^{-7}$
cis			$2.5 \cdot 10^{-6}$ Pa
trans			$1.5 \cdot 10^{-6}$ „
Fenfluthrin	6.2	0.002 found	$1.5 \cdot 10^{-5}$ mbar
Tefluthrin	6.2	0.002 (found)	$2,5 \cdot 10^{-5}$ Pa
Fenvalerate	6.8	0.01	$1 \cdot 10^{-9}$ mm
Cyhalothrin	6.6	0.05	$1 \cdot 10^{-7}$
	(7.0)	0.005 (found)	$1.4 \cdot 10^{-9}$
Fluvalinate	6.4	0.01	$3 \cdot 10^{-10}$
Deltamethrin	6.4	0.01	$1 \cdot 10^{-9}$
	(6.2)	0.00002 (found)	$2 \cdot 10^{-6}$ Pa
Cyfluthrin	6.3	0.1	$3 \cdot 10^{-8}$
		0.001 (found)	
Cypermethrin	6.2	0.1	$2 \cdot 10^{-8}$
Fenpropathrin	6.0	0.2	$1 \cdot 10^{-6}$
Pyrethrin I	5.9	0.2	$2 \cdot 10^{-5}$
Resmethrin	5.9	0.2	$9 \cdot 10^{-2}$
Fenpyrithrin	5.6	0.7	$1 \cdot 10^{-8}$
Cinerin I	5.6	0.6	$4 \cdot 10^{-5}$
Kadethrin	5.4	1	$4 \cdot 10^{-8}$
S 3243	5.4	0.4	$4 \cdot 10^{-4}$
Kikuthrin	5.0	3	$6 \cdot 10^{-4}$
Allethrin	5.0	2	$7 \cdot 10^{-5}$
Barthrin	4.9	3	$1 \cdot 10^{-5}$
Furamethrin	4.5	9	$1 \cdot 10^{-3}$
Pyrethrin II	4.3	0.2	$2 \cdot 10^{-5}$
Terallethrin	4.3	15	$2 \cdot 10^{-4}$
Tetramethrin	4.2	5	$1 \cdot 10^{-7}$
		2–4 (found)	

volatile components, are of interest. This is accomplished by the direct evaporation of a pyrethroid containing solid support, flamelessly heated in a controlled manner. This can be done [456, 457] by a pulp or cellulose sheet, impregnated with a pyrethroid, optionally stabilized by an antioxidant like di-tert.-butylphenol [458]-derivatives. Specially designed miniature electrical heaters are on the market as a controlled heating source. The heat of an electric bulb, however, will do just as well [459]. In the table 84 some physical data are listed which are of technical interest for the use of pyrethroids.

4 Action of Pyrethroids Against Non-Target Organisms

4.1 Ecotoxicology

4.1.1 Effects on Beneficial Arthropods

Where there are many insects, there are usually arthropod-predators preying on them. During the development of pyrethroids suitable for agricultural application, the impact on the important predatory insects, mites and other beneficial arthropods had to be investigated. Although the second generation-pyrethroids were in many cases not so good as miticides, the predatory phytoseid mites like *Amplyseius fallacis* and *Zetzalia mali* are affected somewhat by certain pyrethroids and are much more sensitive than their prey [463–466]. The important predator in cotton, *Amplyseius gossypii* was most heavily affected by oviposition disturbance [467].

Deltamethrin, cypermethrin, fenpropathrin and fluvalinate are very toxic over long periods to the predatory mite *Typhodromum pyri*. Sublethal doses damaged the eggs [468]. In Tse-tse eradication program with permethrin, cypermethrin and deltamethrin a spider species was the most affected arthropod [469] after the fly. In other field investigations (7,5 g/ha) with deltamethrin spiders were strongly affected, while the population of some other beneficial carabid-insects remained intact [470]. A 24-monthfield-study of the effects of summer and autumn spray application of Cypermethrin in winter wheat against the virus transmitting grain aphid revealed no change in number and diversity of beneficial arthropods [471]. The impact of pyrethroids on non-target epigeal insects depends of course on their seasonal occurrence. Spiders are especially sensitive to deltamethrin [472].

The parasite wasp *Eucarsia formosa*, released in greenhouses to combat whitefly is too sensitive to deltamethrin to allow a concomitant treatment against excessive outbreaks of whiteflies [473]. On the other hand, for example, the predatory wasp *Microplits croceipes* showed a more than 10 fold higher tolerance on LD_{50} base to the investigated pyrethroids fenvalerate, flucythrinate and permethrin than the organophosphates applied in their habitat. From this and the lower application rate of these pyrethroids ensures a safety factor for this parasite, resulting in an additional control of crop damaging pests [474].

Another investigated case is *Menochilus sexmaculatus* on eggplants, preying on aphids, which is killed by 20 g deltamethrin/ha and other pyrethroids [475].

Yet, there are many cases, where there is a sufficient pest-predator selectivity by pyrethroids, as for the predator *Toxorhynchites nitilus* [476] on *Aedes aegypti* and for *Campoletis sonorensis* [477] preying on budworms, and *Chrysopa carnea* on *Heliothis virescens* [478]. The latter selectivity is reflected also in the neuropharmacological studies on nerve preparations; specific binding of cis-permethrin was lower in *Chrysopa* [193]. In addition these insects possess a unique very active cis-cypermethrin-esterase [340]. Selectivity of pyrethroid action in a prey/predator systems can also be accomplished by specific formulation in cyclodextrin [421] or in a polymeric carrier [479], by which trick the toxicant is only released on digestion in the stomach of a leaf-eating insect.

It was shown by careful topical application tests, that the selectivity of a pyrethroid between a certain prey and predators is dependent on structures as listed in the following Table 85 [480]:

Table 85. Relative toxicity of pyrethroids to the pest *Ephestia kuehniella* and its predator *Venturia canescens*

Pyrethroid		LD_{50} (µg/insect.)		Selectivity ratio $\frac{Predator}{Prey}$
		Prey *Ephestia*	Predator *Venturia*	
Kadethrin		0.07	10	145
Bioallethrin		0.11	11	97
Flucythrinate		0.026	1.8	69
		0.16	8.0	50
	1R cis	0.08	2.7	34
	1R trans	0.20	5.2	26
1R cis, S-Cyhalothrin		0.004	0.097	27
1R cis-Permethrin		0.10	1.4	14
1R trans-Permethrin		0.16	2.3	14
Deltamethrin		0.032	0.39	12
Bioresmethrin		0.25	2.4	10
Fenvalerate		0.23	1.9	8
(±) c/t		2.4	5.5	2.3

Alphamethrin was found to be less toxic to predators in oil seed rape than the standard organophosphates. In winter wheat no long-term effects on entomophagus insects were observed [481]. 5 beneficial species (*Pterostichus cupreus, Stetorus punctillum, Podiscus maculiventris, Chrysopa carnea, Habrobracon hebetor*) were tested for speed of action of 7 important pyrethroids at an application concentration of 0,01%. Most of them had a 100% knock-down effect within 30 minutes, depending on structure, species and developmental stage [482]. Deltamethrin has a particularly fast action. Fenvalerate and cyfluthrin did not show strong knock-down effects on *Chrysopa* larvae. The commercial knock-down agent against flies, tetramethrin, did not bring about knock-down in *Chrysopa* imagos.

The sensitivity of the lady-bird to recommended rates of pyrethroids is low enough to allow a survival of the population [463] after pyrethroid application in the field.

A selectivity problem with important economic consequences is how to spare the foraging honey bees while combatting insect pests with pyrethroids. As with predators, there is, in the first place, no general rule, since bees are just insects with a different way of life. However, the toxic effects against bees depend on several factors, like place and time of application during the day or in the season, and the structure of the pyrethroid used, which may cause completely different intrinsic activities against bees in comparison to chewing or sucking insects. For example deltamethrin being the most toxic commercial pyrethroid by contact is less toxic to bees than permethrin [491a] and bioresmethrin, which are very toxic to bees. The latter is essentially non-toxic to warm blooded animal [484, 485]. More important than this more or less different intrinsic efficacy of most of pyrethroids against bees in the laboratory is the surprising selectivity in the field under recommended commercial-use conditions [486, 487, 488] as found in several countries and climatic conditions. Fenvalerate, applied at a rate of 110 g/ha in flowering alfalfa in the morning, caused no damage or casualties in beehives next to the field. Bees were repelled for a day [489]. Alphamethrin applied at 20 g/ha on rape and flowering mustard, and 1200 g/ha phosalone, the acknowledged standard low-beetoxic insecticide were both not directly toxic to bees. For a few hours strong repellent effects [490, 491, 491a] were seen with alphamethrin.

Field experiments in Sweden, applying 44 g cypermethrin/ha and 7,5 g/ha deltamethrin in flowering rape early in the morning revealed no serious adverse effect in bees and hives except for the transient repellency [492]. The pollen initially collected in reduced amounts a few hours after treatment contained 0,2 ppm (2 h after) and 0,05 ppm cypermethrin (34 h after treatment). Despite the lower rate of applied deltamethrin the collected pollen contained 0,3 ppm (12 h after) and 0,16 ppm (86 h after treatment). In long-term field tests in wheat deltamethrin (12,5 g/ha) was repellent to bees [484, 493, 494, 495) despite of the fact, that the honeydew of the grain aphid is a very attractive food for bees. Trials in Switzerland with permethrin, flucythrinate, cypermethrin, deltamethrin, and fenvalerate led to the classification for the investigated pyrethroids as 'not dangerous to bees' [496]. Fenvalerate at recommended rates proved to be much less toxic to bees than phosalone.

Glass-house-tests with commercial cypermethrin, however, showed again the intrinsic high toxicity with rapid action on bees. However, bees which are only knocked-down, usually recover within a few hours. Older bees are less sensitive. In the field, bees were repelled and moreover they are not attracted by the smell of flowers under the influence of cypermethrin. After a while, bees learn to avoid pyrethroid contaminated crops [497].

Natural pyrethrins [498] and flumethrin at 1 ppm [497a] are even recommended for selectively killing the Varoa mite, parasitizing bees and endangering larger bee-populations in Europe increasingly these days.

An expanding problem in beekeeping in the American hemisphere is the spread of hybrid specimens of the larger European honeybee and the less diligent, but much more aggressive African bee. This africanized killer-bee is only half as susceptible to permethrin, showing a shallower dose-response-curve [499].

The intrinsic high susceptibility of honeybees to pyrethroids is not due to low detoxication capacity. All enzymes are active. Sublethal permethrin however, stimulated the glutathione transferase [500], which is so important in metabolizing xenobiotic electrophilic compounds. Injected sublethal doses of deltamethrin caused large variations in the fatty ester content in hemocoel possibly due to the release of neurohormones. A great variation in gut alkaline phosphatase was also observed [501].

In comparison to the application of the classical insecticides in cotton e.g., carbaryl (1.25 kg/ha), treatment with deltamethrin (10 g/ha) not only resulted in much better crop protection, but also caused less disturbance to arthropod diversity [502]. Similarly, 500 g/ha dimethoate in winter wheat against aphids caused heavy casualties in beehives, while 15 g cyhalothrin had only a repellent effect on bees, for more than 24 hours despite its high acute toxicity to bees in the laboratory [503].

For a more detailed review on effects of pyrethroids on non-target organisms see J. R. Hill (1985, 293 references) [504].

4.1.2 Action of Pyrethroid on Aquatic Organisms

The intrinsic toxicity of pyrethroids to fish in clear standardized water, as tested in laboratories under more or less generally agreed upon standardized guide-lines is usually very high [505, 506] (Table 86). The toxicity range of the bulk of the trade pyrethroids, presently of importance is between 0.5 to 30 ppb.

Table 86. Toxicity of some pyrethroids to aquatic species in pure water [506]

Compound	LC_{50}/(lethal threshold) ppb				Exposure time
	Salmon	lobster	shrimp	trout	
Permethrin a.i.	12	0.73	0.13		96 h
				135	24 h
formulated				61	24 h
1R cis a.i.		0.4			
Cypermethrin a.i.	2	0.04	0.01		96 h
				55	24 h
formulated				11	24 h
Fenvalerate a.i.	1.2	0.14	0.04		96 h
				76	24 h
formulated				21	24 h

These values are sometimes hard to reproduce because of the questionability of figures obtained by trace analysis of pyrethroids in aqueous systems. This class of very lipophilic chemicals have an extreme tendency in the absence of surfactants to separate from the aqueous phase onto any surface [829], where they stick tenaciously to and are even hard to rinse off with organic solvents. This may easily cause errors of several orders of magnitude by using conventional dilution techniques in the range of ecologically relevant concentrations (see also p. 158). In 1982 the threshold of detectability of fenvalerate was 0.15 ppb [506].

Fish toxicity of pyrethroids is dependent on temperature as found in insects and nerve preparations. In trout an LC_{50} was 0.6 ppb at 5 °C, 6 ppb at 20 °C [506a]. Toxicity goes with an increase of Na and protein level in the serum [506b]. Pyrethroids in clear and cold water are certainly dangerous for trout and salmon [506].

Experimental pyrethroids, a great many from Japanese companies, are claimed to be much less toxic to fish, a basic requirement for application against insects pests in rice. A fish toxicity of 2 ppm is borderline for acceptance of an insecticide for use in Japanese rice fields. These compounds are closely related variations of highly toxic ones, mostly having lower insecticidal efficacy, but also some very different structures.

Some examples for low fish toxic pyrethroids are shown next:

[507, 508]

$x =$ Hal, OCH_3, OCF_3, CH_3
NO_2, $OC_6H_4CF_3$

(see collection of patents Vol. 5 of this senès)

[509]   [62]

[62]

[510]

Particularly non-toxic to fish is the non-ester deviation of pyrethroid structure, the hydrocarbon-pyrethroid MTI 800, having a toxic concentration >40 ppm [62] in combination with high toxicity to rice-pests.

However, the high toxicity observed for other pyrethroides in pure water to fish, whose bloodstream exposed in the gills is almost in intimate contact with the toxicant in the water, is not found under natural conditions in ponds, lakes, streams and the like. Standardized pond experiments with higher than recommended doses indicated the adsorption of the surface applied pyrethroid on plant parts. Only 10% of the pyrethroid was found in subsurface water, where fast adsorption to suspended sediment-particles was observed. Fish were not affected. This essential nontoxicity of deltamethrin to fish and aquatic species under practical conditions was confirmed by extended trials in rice-fields and aquatic systems for 8 year [511].

The impact on aquatic herbivorous insects, however, led to an increase of algae within 2–6 weeks [512]. On the other side, this reduction in aquatic arthropod abundance after contamination of water with pyrethroids like cypermethrin is a hazard to fish population [513]. They may starve as a secondary result [506] and malnutrition is then the cause for a lack of fecundity and a decline of the fish population. Alphamethrin, at 3 to 500 g/ha, ranging from about 1/4 up to 40 times

Table 87. Comparison of ranks of polarity, speed of action and toxicity (Table 88) of pyrethroids to leech and locust [171]

Rel. polarity[a]	compound	
−1.09	A =	1R cis E Kadethrin
−0.57	B =	± cis/trans Cypermethrin
−0.56	C =	1R cis α S Deltamethrin
−0.56	D =	1R t, S Bioallethrin
−0.42	E =	(±) Fenvalerate
−0.07	F =	1R cis Cismethrin
0	G =	1R t Bioresmethrin
0	H =	± c/t Permethrin
0.27	J =	1R t Bioethanomethrin
0.30	K =	1R c Decyanodelta-methrin

[a] ΔR$_M$ relative to G = Bioresmethrin

Table 88.

Decreasing rank of activity

Speed of action knock-down		Toxicity	
Leech[b] Compound	Locust[c] Compound	Leech Compound	Locust[d] Compound
A (2)	C (0,43)	D[e]	C (33)
D (4)	D (0.37)	A	K (8)
F (12)	H (0.36)	S	H (3)
C (21)	E (0.32)	F, G, J	A (2)
B (23)	F (0.14)	B	G (1)
G 220	J (0.06)	H	D, J (1)
J 376	G 0.	E, K	
E, K (1440)	B (−0.025)		
	K (−0.075)		
	A (−0.4)		

[a] ΔR_M relative to G = Bioresmethrin;
[b] (minutes at 10 ppm);
[c] (log relative kd to G);
[d] (relative toxicity), G $\equiv$ 1 $\cong$ 4 µg/g locust;
[e] 0.1 ppm; the others are at least 10 × less toxic; threshold 10^{-8} M

the recommended practical rate, led to aqueous concentrations from 0.2 to 30 µg/l. Toxic effects were observed with 2 to 5 µg/l (2–5 ppb) in pond water, which corresponded to higher than practical rates in crop protection [514]. Similar studies [515] with 10 g/ha labeled deltamethrin led to initial concentrations of 1.3 to 2.5 ppb, which, with a halflife of 2–4 hours, soon vanished. This compound was rapidly adsorbed, mostly by sediment-particles in the pond in addition to uptake by plants and evaporation into the air. Although the fish (the fathead minnow) accumulated the labeled deltamethrin by a factor of 250–900, no mortality could be observed.

This rapid adsorption of pyrethroids is technically used to clear the effluent from wool treatment plants by adding clay powder to the water and precipitating this clay later on [516]. Bioaccumulation, in trout by a factor of 1000, in other species 2000 to 4000, is much less than predicted from the partitioning coefficient of 10^6 [517].

Pyrethroids are metabolized in fish. However, the hydrolysis of trans permethrin occurs 180 times slower than in mice [519]. The halflife of esfenvalerate in carp is 5 days [506]. In reality there is no danger of bioaccumulation, since the rate of adsorption on sediments is much higher. These adsorbates are nevertheless toxic to mayfly larvae [506]. Degradation of pyrethroids in such adsorbents is slower than under atmospheric conditions [506], but a high rate of mineralisation to CO_2 is observed [507a].

In large scale trials in rice fields with cypermethrin no adverse effects on fathead minnow, carp and *Daphnia* could be seen. The same was observed with 13 g/ha deltamethrin in large scale application in the United Kingdom, Spain, Taiwan, Mexiko and Africa [519]. 35 g/ha permethrin applied on a lake area had only a moderate short-term effect on zooplankton in ponds and shallow lakes [520]. At commercially relevant rates of 140 g/ha, however, the benthic fauna and zooplankton were severely affected.

The LC_{50} of deltamethrin against the single cell aquatic microorganism *Paramecium* was found to be 1 ppm, while the no-effect concentration was 0.25 ppm [521].

Toxicity of a number of pyrethroids to frogs depends on structure and stereochemistry of the compound and lies in the range of 0.13–60 mg/kg frog. Trans-isomers are less toxic due to metabolic degradation (1R trans α S cypermethrin: 0.65 mg/kg). Deltamethrin is more toxic to frogs (0.13 mg/kg) by subcutaneous application, producing symptoms [522] similar to the ones observed in rats, which could also be antagonized by the gabaergic compound diazepam. 1 μM deltamethrin in pure water disrupted osmoregulation in the frog *Leptodactylus ocellatus* [523]. Leeches are also extremely sensitive to the action of pyrethroids [171]. The tables 87/88 shows the comparison to locust.

On the other hand, aquatic and terrestrial molluscs and earthworms are barely or not affected by pyrethroids [266]. For a detailed study of the effects of deltamethrin on sweet water organisms see [838].

4.1.3 Effects of Pyrethroids on Plants and Microorganisms

Generally, pyrethroids are well tolerated by plants. However, there are occasional reports of plant growth regulating activity of pyrethroids, such as dose-dependent stimulation of growth of beans (*Phaseolus vulgaris*) (permethrin 10 ppm) [524], synergism with antilodging quaternary ammonium salts [525], growth regulation by cypothrin [526], inhibition of gibberellin-stimulated growth of rice seedlings [527] by deltamethrin (5 ppm). The trans-ethyl-chrysanthemate inhibits the growth of lettuce seedlings [528a] and cis-chrysanthemolactone is also reported as an inhibitor of plant growth [529]. Maybe, the anion of the acid is responsible for this effect, as found in the growth regulating activity of sodium chrysanthemate [530]. Field studies on cotton with several pyrethroids revealed a stimulation of yield by producing smaller, but more vigorous plants in certain cotton varieties, independent of the protection from insect damage [531]. Experiments in orchards with fenvalerate resulted in a reduction of photosynthesis in pecan leaves [532] without visible negative effect. Soybean seedlings are sensitive to the pyrethroids permethrin, cypermethrin, deltamethrin and fenvalerate [527]. Chlorotic effects can be produced with the non-insecticidal RαR'-isomer of fenvalerate [533]. Pyrethrum induces flowering [534]. Certain herbicidal effects can be produced with a compound structurally very close to the only weakly insecticidal pyrethroids, permethric acid pyridylmethylamid [535]. Allethrin was found to be a strong inhibitor of the hill reaction in photosynthetic electron transport [536] without effect on atrazin binding [537]. Deltamethrin with 0.5–2 ppm caused induction of chromosome and chromatide breaks in the root meristem of *Allium cepha* due to disturbance in the mitotic spindle [538].

Reported fungicidal effects of pyrethroids [539, 540, 541] are weak and apparently not of great commercial significance. Organisms mentioned are *Aerobacter aerogens, Pseudomonas aeruginosa, Botrytis cinerea* and *Fusarium*. Permethrin is supposedly more active against *Rhizoctonia solani* than cypermethrin [542]. Pyrethrum inhibits production of the mycotoxine patulin in a penicillium strain [543].

While permethrin had no effect on green algae, it was toxic (EC_{50} 5 ppm) to *Cyanobacterium anabaena* [544]. Bactericidal effects are also claimed for a series of highly insecticidal pyrethroids related to RU 3956 A [540]. Furthermore a number of 3-phenoxy-benzyl-esters (permethrin, cypermethrin, deltamethrin, fenvalerate, fenpropathrin) show minor transient adverse effects on soil inhabilitating microorganisms (bacteria and fungi) with 5 µg/g sandy loam soil [545].

However, the soil microorganisms degrade e.g. cypermethrin quite rapidly. The 1S trans isomer is degraded faster than the 1R trans ester, while the 1R cis ester is more labile than the 1S cis isomer [545a]. The degradation of all isomers is completed after 4 weeks [545b]. Oxidative and hydrolytic processes are involved [545c]. A recent patent application [546] claims the usefulness of pyrethroids for the control of human protozoal infections, particularly against *Plasmodium falciparum*, responsible for malaria [546a].

4.2 Toxicology in Warmblooded Animals

4.2.1 Effect of Pyrethroids on Rats, Mice, Birds

To assess the toxicological risk of commercial pyrethroid application on crops for humans and domestic animals, many investigations in the model animals rat and mice have been undertaken. It was soon found out, that the modern commercial pyrethroids in general possess a very favorable safety ratio rat/insect, making these mostly low toxic insecticides with hitherto unheard of low application rates a major breakthrough for larger areas of modern agriculture (besides their very favorable application costs per crop and per season in comparison to older standards). Because these compounds usually have a low acute toxicity, they could be administered to experimental animals at high doses, whereupon according to Paracelsus law novel toxicological symptoms in rat, mice and other animal were produced upon determining the LD_{50} and were thoroughly analyzed. A large number of publications bear witness to these investigations, which made the pyrethroids to one of the best studied classes of insecticides in the history of plant protection. Higher toxic compounds, which have been in use for many years, are less rewarding in terms of toxicology.

The range of toxicity of experimental and commercial highly-insecticidal pyrethroids to rats or mice varies considerably from highly toxic (below 10 mg/kg p.o.) to essentially untoxic (more than 15000 mg/kg p.o. or i.p). [125]. Toxicity data for commercial pyrethroids are listed in table 89 [547–550]. The determined LD_{50} for a given pyrethroid depends on the type of carrier of the formulation (Table 90) [551], sometimes on the concentration of the pyrethroid therein, on the species of animal, state of animal, (sex, fed or unfed, age) as well as on the stereochemistry of the isomer of the pyrethroid in question, and most important, on the way the pyrethroid is administered (per os, intraperitoneal, intravenous, intracerebral or topical) (Tables 91, 92). As to the dependence of toxicity on concentration, for alphamethrin e.g.

Table 89. Toxicological data of important Pyrethroids

	LD₅₀ mg/kg, acute oral						Fish μg/l = ppb 4 d	Honey bee ng/bee	No-effect-level mg/kg/day		
	♂ Rat	♀ Rat	Mouse	Dog	Domestic fowl	Quail			Animal	dose	observation time
Allethrin	1100	685				2030	Steelhead trout 17				
Alphamethrin		60					Channel catfish 30				
Bifenthrin		55					Rainbow-trout 0.15				
Bioallethrin	784	1545			>5000		Rainbow-trout 10 / Bluegill 33		Rat	1000	3 months
Etofenprox			>20000				(TLm 48: 5000)				
Bioresmethrin	7000–8000				>10000		Guppy 0.5	3	Dog / Rat	>500 / 1200	3 months / 3 months
Cyfluthrin	500–1200		300–600		5000	>5000	Rainbow-trout 0.6		Rat / Dog	300 / 65	3 months / 6 months
Cyhalothrin	243	144	40–50		>2000		Carp >10				
Cypermethrin	250–4123		140–800				Brown trout 2 / Rainbow-trout 25		Dog / Rat	>300 / 100	3 months / 2 years
Deltamethrin	135–5000			>300	>4000	>10000	Rainbow-trout 1–10	50	Rat / Mouse	2 / 12	2 years / 2 years
Tralomethrin	90–3000			>500		>2500	Rainbow-trout 1.6 / Bluegill 4.3		Rat / Dog	6 / 1	3 months / 3 months
Fenpropathrin	164–107				1100	1500–2000	Bluegill 1.9				
Fenfluthrin	85–120		120–160		>2500	>1600			Rat	250	3 months
Fenvalerate	450			Sheep: >2500	>2500		Rainbow-trout 3.6		Dog / Rat	200 / 250	3 months / 2 years
Flucythrinate	81	67	76		>2500	2700	Rainbow-trout 0.3 / Bluegill 0.9 / Sunfish 2.7				
Fluvalinate	6300		690		>3000	2500	Trout 14	80			
Kadethrin	1320–650		540–2700	1000			Rainbow-trout 0.5		Rabbit	25	3 months
Permethrin	430–4000				>3000	>1300	Trout 17		Rat	200	2 years
Phenothrin	>10000					>2500	verytoxic		Rat	2500	6 months
Pyrethrine	580–900										
Tefluthrin	22–35										
Tetramethrin	>5000				>1000	>1000	Bluegill 21		Rat	1500	6 months

a 5 times higher oral toxicity was found in corn oil as carrier at a 4 times lower concentration [551a], possibly due to a faster intestinal excretion or vomiting at higher concentration, thereby removal of the active ingredient.

More lipophilic carriers increase toxicity (Table 90) [574a], while the more polare ones produce a much more favorable toxicity (Table 90) [574b].

Table 90. Acute oral toxicity of some pyrethroids depending on the carrier used

Pyrethroid	Carrier	LD$_{50}$		Lit.
		Rat	Mouse	
Cyfluthrin	PEG 400	650 ♂	610 ♀	
		1200 ♀		
	NMP	500–1000 ♂	610 ♀	
	Isododecan	500 ♂		[550]
	Xylene	500 ♂		
	DMSO	400 ♂		
	Acetone/peanut oil	250 ♂		
	Cremophor dest.	15–25 ♂	< 100 ♀	
Deltamethrin	Sesam oil	128 ♂	33 ♂	
		140 ♀	34 ♀	
	PEG 200	67 ♂	21 ♂	[547]
		86 ♀	19 ♀	[548]
	Peanut oil	52 ♂		
		31 ♀		
	Gum arab		5500 ♂	[549]
			3500 ♀	
Fenvalerate	PEG/H$_2$O	> 3200	1200	
	DMSO	450	200–300	
Permethrin	Water	2950 ♂	> 4000	
		> 4000 ♀		
	DMSO	1500 ♂	250–500	
		1000 ♀		
	Corn oil	430–500 ♂	650 ♂	
		470 ♀	540 ♀	
Cypermethrin	Water		780 ♀	
	DMSO		138	
	Corn oil		82	
Resmethrin	Polyethyleneglycol	> 3000 ♀	> 3200 ♀	
	DMSO	1347 ♀		
	PEG 400	2000 ♀		
	Corn oil	> 5000 ♂	690 ♂	
Allethrin		920	480	
	Corn oil	2430 ♂	500	
		720 ♀		

Table 91. LD_{50} of Pyrethroids by different application [559a, 560]

Pyrethroid (in Glycerol)	Stereochemistry	LD_{50} mg/kg ♀ rat	
		Intravenous	Oral
	Pyrethrin I 1R trans; S	2–5	100–1500
	Pyrethrin II 1R trans, E; S	0,5	100–1500
	Bioallethrin 1R trans; S	<10	100–1500
	Bioresmethrin 1R trans	340	>4000
	1R trans, Z	10–200	100–1500
	Bioethanomethrin 1R trans	<10	<100
	Cismethrin 1R cis	6–7	<100
		4–5	100–1500
	1R trans 1R cis	<10 0.5	100–1500 14

Table 91. (continued)

Pyrethroid (in Glycerol)	Stereochemistry	LD$_{50}$ mg/kg ♀ rat	
		Intravenous	Oral
(structure: dichlorovinyl cyclopropane carboxylate furylmethyl benzyl)	1R trans	10–200	100–1500
	1R cis	1.4–2.8	13
(structure: dimethylvinyl cyclopropane carboxylate phenoxybenzyl)	1R trans	> 200	10000
(structure: dichlorovinyl cyclopropane carboxylate phenoxybenzyl)	1R trans	> 200	> 1500
	1R cis	> 200	> 1500
(structure: α-cyano phenoxybenzyl ester)	1R trans; RS	10–200	100–1500
	1R cis; RS	< 10	100–1500
(structure: dibromovinyl cyclopropane carboxylate α-cyano phenoxybenzyl)	1R cis; S	2–3	100–1500

Although the cis-isomers are less readily resorbed from the stomach than the trans-isomers [574], the cis-isomers are generally more toxic than trans-isomers as shown by cerebral application; Table 93. There is no linear relationship of cis- and trans-isomer content in permethrin and determined LD$_{50}$ in rat or mouse [551] (Table 94). Compared to insect, the difference between the two stereoisomers is more pronounced in rat than in mosquito [553] as shown for the resmethrin-isomers [554] (Table 95).

Beside the differing LD$_{50}$, pyrethroids cause strikingly, different symptoms in animals, depending on the presence or absence of a small structural feature as exemplified best by cismethrin and deltamethrin [557, 558, 565]. These dramatic differences in toxicity are not explained by higher metabolism, but certainly by a lack of affinity to a specific binding site in the nervous systems involved [555, 556]. This small structural difference is the α-cyano-group in the benzyl position of the alcohol component of these esters.

Two groups of symptoms divide the pyrethroids in two types, type-I- and type-II-pyrethroids, the former lacking the cyano group. There are compounds which belong to either or neither one [559]. This subdivision holds for action in all other animal, from cockroach to frog, to rat and moreover in nerve preparations in vitro. Type-I-pyrethroids produce the T-syndrome in rats: aggressive sparring, sensitivity to external stimuli, production of repetitive discharges in the muscle, progression to gross, whole-body tremors and prostration.

Table 92. Acute toxic effects against warmblooded animal in dependence of mode of application

Permethrin	Formulation	Acute toxic effect mg/kg								
		Sheep		Cattle		dog		cat	rat
		i.v.	Ruminal	Ruminal	Abomasal	i.v.	oral	oral	oral
25:75	Corn oil	300 Minor Effect	800 Minor Effect	400 Tremors 800 Lethal	800 No Effect	200 No Effect 400 lethal	16000 No Effect	200 Lethal	3610–4670
	Neat						500 6 months No Effect		20000
	40% DMSO								>8000

Table 93. Intracerebral toxicity to mice [561]

Pyrethroid	Stereochemistry	LD_{50} µg/g brain	Symptom
	1R cis	7	
	1R trans	14	
	1R cis	40	
	1R trans	> 8600	
			type I tremor convulsions
x = CH$_3$ y = H — 1R cis	> 4300	
	1R trans	> 8600	
x = Cl y = H	1R cis	11	
	1R trans	> 860	
x = Cl y = CN	1R cis, RS	0.6	
	1R trans, RS	1.6	
x = CH$_3$ y = CN	1R cis, RS	4	
	1R trans, RS	12	
	±	6	type II Choreoathetosis convulsions salivation
	S, αS	1	

Table 94. Toxicity of permethrin — isomeric mixtures

Permethrin % isomer				LD_{50} mg/kg		
				Formulation	Rat ♀ oral	Mouse ♂ oral
1R cis	1S cis	1R trans	1S trans			
	80		20	40% in cornoil	200– 250	
	60		40		400– 500	
	50		50		730–1360	
	40		60		1000–1590	490
	30		70		1250–2250	
	20		80		6000	
		100				3100
			100			> 5000
100						107
	100					> 500

Table 95. Different toxicity of two typical pyrethroid isomers against rat and mosquitoe

	LD_{50} rat p.o.	i.v.	LD_{50} mosquito topical	Therapeutic index ratio of individual toxic dose p.o. rat/mosquita
1R cis Cismethrin	32 mg/kg ($\sim$5 mg/animal)	5 mg/kg ($\sim$1 mg/animal)	0,64 ng/animal	10^5
1R trans Bioresmethrin	8000 mg/kg ($\sim$1200 mg/ animal)	500 mg/kg ($\sim$80 mg/ animal)	0,95 ng/animal	10^9

Type-II produce the *CS*-syndrome: pawing, burrowing behaviour, head bobbing, salivation, coarse tremor in sinuous writhing and choreoathetosis, clonic seizures, prolonged extra discharges in the trigeminal motor nucleus [563], besides repetitive discharges in muscle after external stimulation.

Neither a clear cut relation of structure with symptoms could be deduced, nor a straight forward explanation of the origin of this very complex toxicological response in central and peripheral parts of the nervous system, even after topical application by infusion of deltamethrin in certain parts of the neuraxis [564]. In addition, pyrethroids also have a spinal aspect of action [565, 566].

Cismethrin and deltamethrin dramatically facilitate spontaneous firing in ventral roots and spinal interneurons. Maybe, the heterogeneity of Na-channels is involved [567]. The simpler symptoms of cismethrin can be explained by action on the trigeminal reflex system [558]. The T-syndrome is electrophysiologically accompanied in the peripheral nervous system by enhanced sensitivity to external stimuli, by spontaneous activity and by repetitive firing of axones. The CS-syndrome is characterized by spike and wave discharges.

Table 96 shows the different types of syndromes caused by different pyrethroids.

Table 96. Intravenous toxicity and poisoning syndromes of closely related pyrethroids in rats [562]

Compound	Stereo chemistry	Toxicity mg/kg	Type of syndrom
X = CH$_3$	1R cis	6–7	T
	1R trans	340	T
X = F	1R cis	0,3–0,7	T
	1R trans	1,4–2,8	T
	$\pm$	2–5	T

Table 96. (continued)

Compound	Stereo chemistry	Toxicity mg/kg	Type of syndrom
	±	2,5	T
X = Br X = H	1R cis	> 30	T
X = Br Y = CN	1R cis, S	2–2,6	CS
	1R trans, S	17–25	CS
X = CH₃ Y = H	1R trans	> 570	T
X = CH₃ Y = CN	1R trans, S	17–25	T
	1R cis, S	5	CS
X = F Y = CN	1R cis, S	0,5–0,8	CS
	1R trans, S	5–7	T
	±	50–100	CS
	±	200–250	TS
	±	225–280	TS
	±	400–500	T

T = tremor; C = choreoathetosis; S = salivation

Yet, the onset of different contributions to the CS-symptoms in the sequence salivation, tremor, choreoathetosis could be correlated with the level of i.p. applied deltamethrin in blood and brain, Table 97.

Deltamethrin is rapidly transferred from blood to brain [568], but is also rapidly metabolized in the liver and the blood. There is a sharp threshold concentration of 0,5–1 nMol/g for reversible symptoms [569] in the brain. Only a very small part of the i.p. applied dose actually accumulates in the nerve tissue.

The peak level of deltamethrin in the brain, however, did not correspond to the

Table 97. Symptoms of intoxication of rats in dependence of level of deltamethrin in blood, brain and spinal cord

Symptoms	Level of deltamethrin nM/g, (applied dose mg/kg i.p.) rat			
	In blood		In brain	In spinal cord
No symptoms		(<2,5)		
Salivation	1,5–2,4	(2,5)	0,12 (8)	0,24 (8)
Tremor	2,5–3,8	(5)	0,22 (8)	0,36 (8)
Choreoathetosis	3,6–5	(7,5)	0,38 (8)	0,41 (8)
Death	>5	(>8)		

severity of symptoms [564]. A sublethal dose (2 µg/kg i.v.) of deltamethrin caused an excessive increase in cerebral bloodflow, lactate, glucose and NH_3 in the blood [570] as well as an increase of glucose-metabolism in many areas of the brain [571, 572, 573] such as in the cerebellum, colliculi and hypothalamus during the early stages of poisoning. This happened before spike discharges in the EEG appeared, independent of motor disturbances. In the cerebellum, these two effects are correlated with extrapyramidal motor hyperactivity, which in addition is also connected with elevated c-GMP-levels in this part of the brain [575, 577]. However, the last effects are not directly correlated to deltamethrin levels. The distribution of labeled deltamethrin in various brain tissues during different stages of poisoning was also investigated [579].

In addition to the increased glucose utilization in the brain there is a general tendency for increasing gaba activity and an increased concentration of noradrenaline and cGMP in the cerebellum [568, 570], while cAMP is unchanged. The increase of glucose turnover to CO_2 caused by sublethal doses of deltamethrin was also observed in the locust. Reasons for this may be the inhibition of oxidating enzymes in the pentose cycle e.g. glucose-6-phosphate dehydrogenase, or stimulation of enzymes in the Krebs cycle by this pyrethroid [570a]. An effect of cypermethrin on blood is the hemoglobinization of red blood cells and the induction of lymphopenia together with neutrophilic leukocytosis, reduced serum albumin and increased serum globin [580]. Livers of rats surviving a LD_{50} of cypermethrin were only slightly enlarged. The non-toxicity of bioresmethrin in the rat brain is not due to enhanced metabolization in liver, instead the non-interaction with the binding site of other pyrethroids seems to be the case [581].

A sublethal oral dose (50 mg/kg) of deltamethrin reduces, and resmethrin inhibits the oxytocine and vasopressine release and content in the rat hypophysis, in the same way as accomplished by calcium channel blockers verapamil or the Na-channel blocker TTX [579].

This is explained by the initial prolongation of the Na-current, followed by a decrease of amplitude of the action-potential, then decreased opening of the Ca-channel. Thus the lack of internal Ca reduces the exocytosis of neurohormons.

High doses, 2000 ppm of permethrin, cypermethrin, fenvalerate and deltamethrin, in the food which can be fed to birds (e.g. quail) caused an increase of the oxidizing hepatic enzymes cytochrome P 450, aldrin-epoxidase, ethoxiresorufin-dealkylase and NADPH-cytochrom-reductase. On the other hand, a decrease of the other oxidizing hepatic enzymes 7-ethoxycoumarin-dealkylase and anilin-hydroxylase was observed [582].

Sublethal doses of deltamethrin (2 mg/kg i.p., 10 mg/kg/day or 200 ppm in the food p.o.) and permethrin (50 mg/kg/day p.o.) in chronic studies with rats resulted in only weak inductions of liver microsomal cytochrome P 450 and NADPH-cytochrome-reductase [583, 584], thus habituating the rat to the toxin. In addition, it caused a prolonged period of increased nerve excitability after external stimulation [585](much longer than with cismethrin), a decreased sensitivity to pain [586] and an adverse effect on learning, inquisitiveness and operant response [587, 588].

Subacute and long term doses of permethrin and cypermethrin cause moderate enlargement of the liver without increase in weight, increased microsomal enzyme activity and smooth endoplasmatic reticulum proliferation. These effects, however, are readily reversible after suspension of the treatment [589].

The quite differently structured tetramethrin also caused enhanced activity of the oxidizing enzymes cytochrom P 450, aminopyrine-demethylase and anilin-hydroxylase [590] and increases hepatic lipids and triglycerides.

One twentieth of the LD_{50} of permethrin decreased the electrocardiogram frequency, but did not show changes in the electroencephalogram and only slight electromyogram modifications [587]. A transient neuromuscular dysfunction was seen one week after administration, while biochemical changes as measured by increased β-glucuronidase and β-galactosidase were seen after 4 weeks of administering subacute doses of permethrin. Cypermethrin and deltamethrin were observed in distal portions of sciatic/posterior tibial nerves. This was not the case with resmethrin [591].

However, there is no direct correlation between the time-course of neuromuscular dysfunction and the observed biochemical changes, suggesting a short term pharmacological effect separate from a more neurotoxic effect in near-lethal concentrations [592]. In chronic studies with subacute doses of permethrin no neurotoxic effects on nerve fibres were seen [593].

Investigations of neuropharmacological effects of pyrethroids on mouse neuroblastoma cells and in some kinds of mouse brain synaptosomes, using radioactive sodium, showed that nano-molar amounts of allethrin and deltamethrin keep Na-channels open, after they have been opened up by Na-channel opening drugs like veratridine or batrachotoxin [354, 596], in that they bind at a different site than other Na-channel-effectors. This prolonged Na-flux into the nerve cell is cut off by the Na-channel-blocker tetrodotoxin (TTX), as found in insect-preparations. However, these experiments failed with Holans DDT-pyrethroid-hybrids [597], which are in many other respects like pyrethroids.

This effect is stereospecific and observed only for certain insecticidal or toxic isomers and is reflected also in stereospecific binding in distinct receptors found in the mouse brain. Specific binding is much more discriminating in mouse cells than in insects. So, the insecticidally very active bioresmethrin and biopermethrin trans-isomers are not recognized by these receptors. However, the cis-isomers are strongly bound. The fastest acting pyrethroid in flies was also the fastest acting in the brain by intracerebral application [578].

Two types of receptors have been identified in mouse brain. One binds non-α-cyanopyrethroids, discriminating 1R cis- and 1R trans- resmethrin and permethrin respectively, but not 1R cis- and 1R trans-ethanomethrin. The α-CN-pyrethroids bind specifically to the other one, without great differentiation between cis- and trans-isomers.

In addition to the established impact on Na-channel-inactivation, leading to prolonged Na-current into the cell and nerve conduction facilitation, there are several hints of the action of pyrethroids on the Cl-channel, as anti-gaba-effectors. However, deltamethrin inhibits Cl-flux into the nerve cell in concentrations much higher than found effective for action on the Na-channel [599]. In fact, deltamethrin and other pyrethroids stimulate the release of gaba [600, 575]. Together with the antidotal effect of the indirect gabaergic effector diazepam, all this supports the likelihood, that in animals, subjected to a large dose of an insecticidal α-cyanopyrethroid, the gaba-effect in combination with the Na-effect may play an important role in the complex syndrome, i.e. they may be potentiating each other.

Whether the observed inhibition of ATPase [601] is a secondary impairment, will be found out in the future. Inotropic concentrations of deltamethrin did not inhibit NaK-ATPase. The contractive effect on guinea pig muscle was brought about indirectly by stimulated release of catecholamines due to facilitated Na-flux [602]. Some only weakly insecticidally active pyrethroids modulated the heart beat [603, 604].

Pyrethroids are even much less toxic to birds than to rats (table 89), although there is a report on relatively high levels of pyrethroids in the brain of chicken after oral administration of a single 15 mg dose, in connection with slow metabolism in that organ [836]. There are reports on a decreased haematocrit [505] and on supression of immune function in chicken, rat and rabbit after application of 1/20 of a LD_{50} [606]. Immune suppression seems to be involved in the inhibitory activity of pyrethroids on the mitogenic responsiveness of murine spleenic lymphocytes to concanavalin A [607]. Decreased immune response in rabbits or rats, immunized with Salmonella typhimurium, was also reported for cypermethrin [608]. It is assumed to be the very first detectable physiological effect in rats after exposure to this compound [608a].

Technical deltamethrin has a repellent action on quail, but the purified compound even attracted the intoxicated quail [609].

Recently permethrin gained importance in technical wood protection as superior substitute for the older compounds like Lindane®. The harmlessness of permethrin to bats living under old wooden belfry roofs protected by permethrin confirmed once more the safety of this highly efficient insecticide to warm blooded animals under practical circumstances [610].

All the poisoning symptoms in warmblooded animals mentioned before are the result of more or less forced pyrethroid uptake in doses several 1000 fold higher than likely to occur during exposure under recommended practical conditions in the field or household.

Apparently, higher doses of pyrethroids may cause many effects. Labeled cypermethrin taken up orally by the rat in 1/100 of the LD_{50} was distributed within 24 hours over the body as shown in figure 17; the bulk to the fat bodies, adrenals, skin and so on. Less than a 0.1% got to the brain [594]. Due to rapid metabolization the label disappears from the body within days.

The only major difference between the sexes is the fact, that the female gonads accumulate much more of the labeled cypermethrin (Fig. 17).

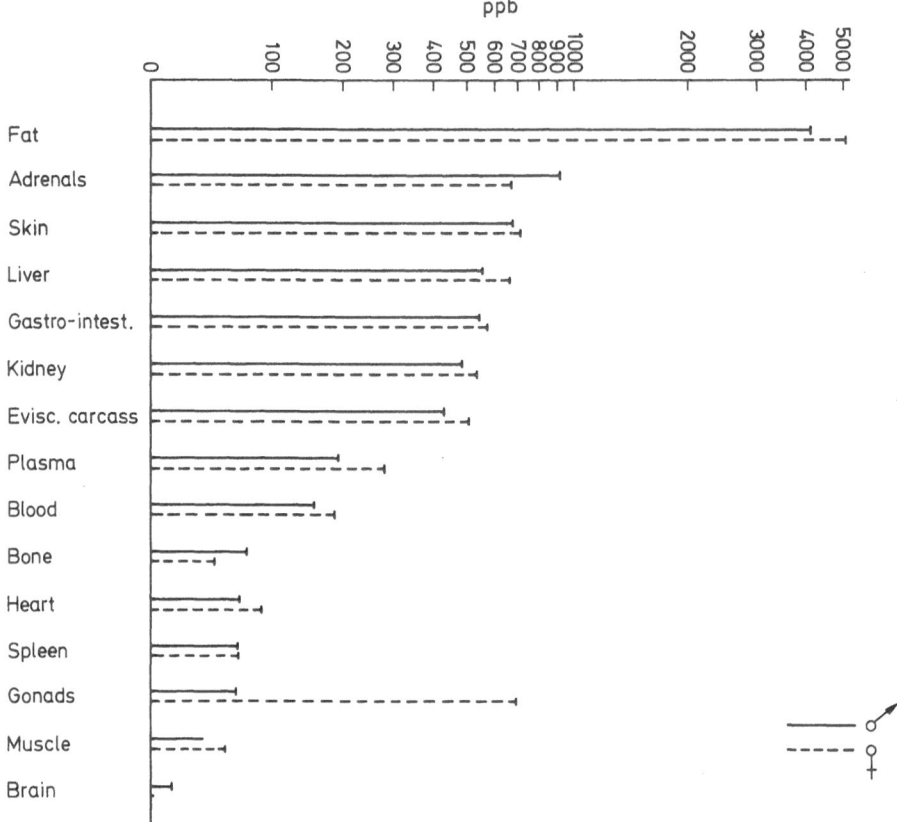

Fig. 17. Mean relative distribution (ppb) of labelled cypermethrin in rat tissue 24 hours after administration of sublethal 2 mg/kg p.o. For a more detailed analysis of mamalian toxicology of pyrethroids see [621a]

4.2.2. Metabolism in Warmbluoded Animal

Rapid metabolism in blood and liver usually is the reason for the very favorable toxicity of pyrethroids. Rate and stereoselectivity of hydrolysis and hydroxylation depend very much on the species [594e]. Only one minor liver esterase out of 7 in the mouse is capable of hydrolyzing fenvalerate very slowly. The main esterase out of six in the rat cleaves it, while all seven in the rabbit hydrolyze this molecule rapidly [578]; preferentially the esters of the R-acid are hydrolyzed [594b].

α-Cyanoesters in the mouse are slower degraded by oxidases and esterases than the primary ester pyrethroids. Depending on structure, 1-R-trans or 1-S-trans esters are preferentially hydrolyzed [576]. While cis-cypermethrin and cis resmethrin is cleaved hydrolytically in the mouse, its ester bond is cleaved involving oxidation in quail [576b, 576c].

Most of the metabolites found so far may be found in all of the species investigated. A shortened summary for a few structures is given in the next scheme for metabolites found in rat and mice [594a, 594b]. Arrows indicate sites of hydroxylation (insertion of oxygen into CH bond), further oxidation and hydrolytic cleavage by esterases and lipases. The thickness denotes the importance of the attack:

The preferential oxidative attack on the cyclopropane moiety occurs at the methyl group in the active isomer which is cis to the ester group [594c]:

1 R cis   1 R trans   1 S trans

However, in the inactive isomer, the trans-methyl group is the one which has the right orientation during to the oxidase for the hydroxylation [594d].

In absence of esteratic cleavage in a modified microsomal system oxidation occurs on both methyl groups. The α-cyano group reduces the hydrolytic and the oxidative metabolic rate as well [594g]. Another noteworthy difference in the metabolism of permethrin and cypermethrin, is the additional aromatic hydroxylation in the 6-position of permethrin, but in the 5-position of cypermethrin and deltamethrin [594e], in addition to the facile hydroxylation in the 4'-position of the alcohol moiety in all cases.

A more detailed study [831] of the microsomal hydrolytic power of different mouse tissues showed a remarcable stereoselectivity in the cleavage of the diastereomers of fenvalerate. Kidney, spleen and brain cells cleaved the non toxic RαS isomer. Liver microsomes hydrolysed the RαS and the RαS isomer much better than the SαR and the toxic SαS isomer. On the other hand. The SαS and the RαR isomer were preferentially hydrolysed by the plasma, leaving the other two isomers for further metabolism. The R isomer of fenvaleric acid was transformed into the cholesteryl ester only in kidney, spleen and liver cells. The most active metabolizing tissue was found in the liver.

For a recent extended discussion of the metabolism of pyrethroids see Leahey (1985, 111 references) [594f].

4.2.3 Effect of Pyrethroids on Humans

An old observation [840] by certain people while using overdoses of pyrethrum and its concentrate in their own vicinity, is a peculiar skin sensation on certain parts of the face and other parts of the body surface, which the author by his own experience could best describe by the expression 'cold burning', a strange paresthetic sensation. This sensation can become unpleasant and may last for a few hours, in certain cases up to more than one day, depending on the structure and amount of pyrethroid in question and the sensitivity of the person, in addition to the type of formulation of the active ingredient. Maybe the very name pyrethroid has its root in the greek word pyr = fire. Some pyrethroids produce no effect, while others are very skin irritating. Apparently, this side-effect prevented the development of certain very promising pyrethroid insecticides and acaricides for commerical use. It is not easy to find a model animal to test this property, although the rabbit ear [611] and the back of the guinea pig [612, 613] are used frequently as a model even for quantitative assessments [613a]. The α-cyanopyrethroids are supposed to have a higher tendency [612] to skin irritation. However, a generalization is not justified.

One investigation [614] found the relative rank order for cutaneous sensation on the human ear lobe, beginning 30 minutes after application, coming to a maximum after 8 hours and disappearing after 24 hours. When the relative efficacy of permethrin in this test equals one, then cypermethrin is also one, fenvalerate is four and flucythrinate is nine [615]. Another study [613] with emulsifiable concentrates indicated the sequence for skin irritation in rats and guinea-pigs to be: deltamethrin > permethrin > cypermethrin. Esters of the fluvalinate acid with α-methyl- or α-methoxycarbonyl-6-phenoxypyridyl-2-methanol [75] are reported to be particularly skin irritating. Not only pyrethroids, but also formulation additives and solvents, or even synergists like piperonylbutoxide, sometimes caused a transient sensation on human skin. By changing to a formulation of deltamethrin in water, this unwanted side-effect could be minimized.

During occupational exposure symptoms of irritation were reported to have occurred also in the upper respiratory tract of workers, spraying fenvalerate and permethrin [616]. Under similar, apparently careless conditions in a chinese factory 2/3 of the humans exposed to about 5–12 mg deltamethrin per m^3 air or 12 to 55 mg fenvalerate per m^3 experienced the irritating effects (cold burning and numbness of the skin). The other third suffered from sneezing and eye-watering. In addition, headache, heartburn and skin spots were somewhat dependent on the time of the year [617].

Cypermethrin proved to be unsuitable as a mothproofing agent for raw wool, since the particels from this treated wool during manufactural handling sometimes caused the above mentioned symptoms.

Human liver esterases hydrolyze both cis- and trans-cypermethrin equally rapidly [615], and faster than the ones from rats, which cleave the trans isomer 10 times faster than the cis isomer. As in other mammals hydrolysis is also the major pathway of metabolism in humans. Within 24 h 80% of the trans isomer and 50% of the cis isomer were excreted [618].

So far as commercial pyrethroides have been thouroughly investigated in many toxicological studies for and during the last 20 years they turned out to be devoid of any acute danger to humans in doses which are applied as insecticides.

Long term chronic studies also revealed no adverse effects due to the exposure to higher doses taken up by mice, rats, dogs, birds, rabbits, guinea pigs, sheep and other warm blooded animals. Daily uptake of LD_{50}-amounts or just sublethal doses by test animals resulted in histological changes in the survivors. In spite of these high doses there was, in general, no significant mutagenic or carcinogenic activity of the commercial pyrethroids in the standard test animal, the mouse. In the beginning, however, the last point needed further extended clarification concerning permethrin [620, 620a]. In certain in-vitro-tests for mutagenicity, like the ames-test, some pyrethroid chrysanthemates produced positive results, whose significance for warm blooded animals remains to be proven in view of conflicting evidence from other assays [621, 622] (see Table 98). May be the epoxides of chrysanthemic acid derivatives (p. 106) are involved.

Epoxides of halovinyl pyrethroids, products of reactions with ozone, are only weak bacterial mutagens [847].

Table 98. Effect of some pyrethroids in in-vitro mutagenicity tests [Lit.]

Pyrethroid	Assay	Effect	Mutagen	
Allethrin	Salmonella, TA 100	+ [623]	+ [625, 626]	[623, 627]
Allethrin-photo-oxydation-product			+	[628]
Bioallethrin	Salmonella typhi-murium reverse mutation	+ [624]	+ [626]	[629]
Tetramethrin	Human amnion cell, Salmonella		+	[623]
Phenothrin			+	[623]
Furamethrin			+	[623]
Resmethrin			+	[623]

A recent review on cases of human pyrethroid intoxications in China is published in [841]

4.2.4 Interaction of Pyrethroids in Vertebrates with Other Drugs

In view of the novel toxicological symptoms produced by the uptake of about 50–5000 mg of pure commercial pyrethroid per kg rat (corresponding to 3–300 g per man) the toxicologists and pharmacologists were confronted with the necessity of looking for antidotes or for potentiating effects by other drugs. Moreover, either class of compounds would shed some more light onto the mode of action of pyrethroids in producing the primary or secoundary effects in mammals and other animals.

The following is a list of reported cases where the combination of drugs with pyrethroids resulted in higher activity.

- Esterase inhibiting phosphoric esters like profenofos, sulprofos, EPN (0,5–5 mg/kg i.p.) enhanced toxicity of trans-cypermethrin in mice 20 fold [630].
- Deltamethrin potentiates the hypnotic action of chloral in mice, stimulates the toxicity and prolongs convulsive seizures of pentetrazol [631] and strychnine, and toxifies tryptamine, presumably by inhibition of monoaminoxidase [632].
- The serotonin blocker methysergit and the serotonin depletor p-chlorphenyl-alanine potentiate the allethrin-induced tremor in mice, which were antagonized by 5-hydroxytrypthamine [633].
- The neurotoxic symptoms of cis-permethrin, trans-permethrin (30 mg/kg) and deltamethrin are potentiated in mice after doping with centrally acting noradrener-gic, nicotine-cholinergic and serotoninergic effectors like clonidine (0,5 mg) [634], chlorpromazin (6 mg), reserpine (5 mg) and mucamylamine (10 mg).

4.2.5 Antidotes Against Pyrethroid Action

Antidotal compounds, which alleviate or abolish the toxic action of pyrethroids are listed in the following.

In insects the inactive 1S cisα R enantiomer of the mirror image of deltamethrin acts as an antagonist to deltamethrin in musca and phaedon by a factor of about 2 [104, 19]. However, this antidotal effect has not yet been reported for mammals.

Sedative benzodiazephins are stimulators of gaba binding, thus enhancing the postsynaptic chloride inward current as counterbalance to nerve facilitation due to prolonged presynaptic Na-inward current. They act as prophylactic partial antidotes to pyrethroids as shown for diazepam and tetrazepam [634]. For instance, diazepam delays the onset of symptoms of the α-cyano-pyrethroids deltamethrin and fenvalerate, but not of permethrin and allethrin on cockroaches and mice [635] (protection by a factor of 6–9).

In frogs, diazepam was more effective than flunitrazepam [522] against del-tamethrin. Tetrazepam with its additional muscle relaxing properties is 3 times more effective on the T-symptoms elicited by fenfluthrin [636], than on cyfluthrin-effects.

In mice, diazepam and the anticonvulsiva clonazepam and Na-valproate could not prevent the onset of deltamethrin induced CS-symptoms [637] in doses which were effective against pentylenetetrazole symptoms.

PK 11195, an antagonist of the peripheral benzodiazepin binding site caused complete reversal of the proconvulsant action of permethrin and deltamethrin [631].

Pretreatment of mice with aminooxyacetic acid or cycloheximide [164-4] also diminished a part of the symptoms of permethrin. Less potent was phenobarbital [635], but still effective against deltamethrin and permethrin. Prophylactic atropine treatment abolished hypertension and bradycardia as the lethal cause produced by allethrin in dog [638]. Blockers of noradrenergic receptors or of norepinephrine synthesis like pentobarbital, chlorpromacine, pentolamine, fenoxibenzamine and reserpine depressed the tremor in mice caused by allethrin. Procaine amide and propanolol are prophylactic preventors of some of the deltamethrin induced symptoms. Similarly, partial success is reported for phenytoin against permethrin [640]. Methacarbamol lowered the death rate and alleviated the motor symptoms after curative treatment of poisoned rats [641]. Russell Uclaf recommends ammobarbital against convulsions caused by poisoning with deltamethrin [645].

Intravenous treatment of rats with the muscle relaxant mephensin shortly before the intravenous injection of cismethrin or deltamethrin, gave a complete protection against all toxicological effects. Also intravenous continous treatment with mephensin after intravenous intoxication with a pyrethroid saved all rats, while the untreated ones died. This lifesaving antidotal action is supposed to be due to the depression of polysynaptic interneurons in the spinal cord [642] by mephensin.

An immediate treatment of rats, poisoned intravenously with an LD_{50} of a pyrethroid, with a cocktail of atropine, diazepam and clomethiazol raised the LD_{50} more than three fold [646].

The clinical symptoms in deltamethrin poisoned dogs can be rapidly moderated with clomethiazol for a short time. Diazepam gives a slower but longer lasting relief [644].

No prophylactic effect on the pyrethroid induced toxic symptoms in mice or rat have been found with α- and β-adrenoreceptor-blockers, gaba mimetics, morphine [633], baclofen, meprobamate, the anticonvulsant phenytoine and muscarinic agonists. The muscarinic antagonist atropine supresses certain symptoms like hypersecretion [644].

Against the persistant cutaneous skin irritations the local anesthetica quinisocaine [266] or benzocaine [646] are reported to be effective. The antioxidant α-tocopherol (vitamin E) [614, 617, 646a] is also useful. It stabilizes liposomal membranes and prevents thereby the liberation of histamine and serotonin from mast cells and tissue cells [614].

By virtue of their Ca sequestering properties gangliosides antagonize the opening of Na channels by pyrethroids [830].

C. Mode of Biological Action of Pyrethroids

5 Poisoning Events of Pyrethroids in Insects

Once an insect has taken up a toxic dose the next steps depend on the temperature and on the type of pyrethroid. Some act more or less fast, knocking down the insect or they give rise to other typical symptoms which prove the neurotoxic action of pyrethroids. In the meantime topically intoxicated flies try by vigorous grooming to remove the toxic substance more or less successfully [648]. The lethal effect occurs much later. The typical stages of poisoning in cockroaches in connection with their accompanying electrophysiological symptoms, measured in live cockroaches at different temperatures and doses, are listed in Table 99 [649].

Table 99. Electrophysical protocol of action of pyrethroid on the free walking cockroach

Temp.	Visible symptoms of poisoning	Electrophys. symptoms at the nerve level	Time after dosing
15 °C Dose 1,78 µg/insect.	Restlessness	— Non stimulated: normal excitation of afferent neurons — Stimulated: a long train of spikes here after	5 min.
	Uncoordination	— Large-amplitude axonal spikes Increasing in frequency independent of stimuli; — Long spontaneous spikes of efferent potentials	40–60 min.
	Prostration	— inconsistent after-discharges after stimulation — Prolonged firing of sensory axons and interneurons after stimulation	1.5–3 h
	Paralysis	After-discharges and nervous activity declining after stimulation	4–8 h
		Response to stimulation blocked, first in cercal nerve, then in ganglion	27 h–45 h
		Evoked response to nerve cord stimulation blocked	4–6 days
32 °C 17 µg/insect.	— Restlessness	As above	Immediately
		— Intense nervous activity, spike trains	

Table 99. (continued)

Temp.	Visible symptoms of poisoning	Electrophys. symptoms at the nerve level	Time after dosing
	− Uncoordination and − Hyperactivity − Tremor	− Abdominal peripheral interneuron after-discharges following stimulation following stimulation	30 min
	− Prostration	− After-discharges following stimulation in ganglion; abdominal after-discharges − Increasing and prolonged firing of cercal afferent and efferent axons.	1.5–3 h
	− Paralysis	− Block of nerve cord	3–4 h
	− Death	− Block of all nervous activity	20–60 h
1.78 µg/insect.	Transient restlessness		Within 15 min.

The marked negativ temperature coefficient varies between different nerves within the insect [649]. Direct excitatory effects on the nerve disappear at lower temperatures [650].

These symptoms originate from neural disorders from many causes, which have been analysed thoroughly to solve the question of central versus peripheral modes of neurotoxic action of pyrethroids.

The strong hyperexcitation at the beginning points to effects in the peripheral nervous system [649]; it could be interpreted for example as a blockage of inhibitory neurons. The central effects at 15 °C (see Table 99) are of secondary nature. The nerve blockage occurs many hours after paralysis and is not the critical factor in causing death [649] in the cockroach.

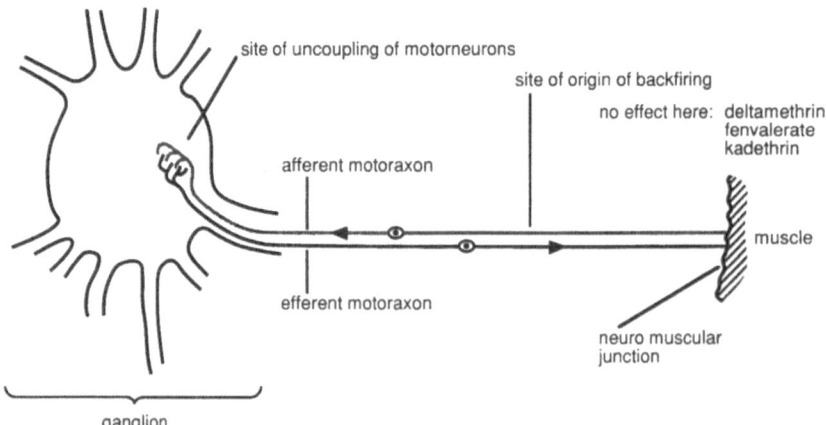

Fig. 18. Pyrethroids as nerve ending poisons. Different sites of action of pyrethroids on the nerve axon

Table 100. Neurophysiological correlates of pyrethroid poisoning in alive flies [652]

Compound	Polarity	LD_{50} mg/fly	Dose applied mg/fly	Neurophysiological effect		
				Uncoupling of moto-neurons in ganglion	Repetitive backfiring of moto-axon to CNS	Transition Temp. from rep. backfire to single potentials
Deltamethrin	−0,56					
(structure: Br, Br, CN, COO, O)		0.003	0.05	+	−	
1 R cis α S *Fenvalerate*						
(structure: Cl, CN, COO, O)		0.04	0.2	+	−	
(±) *Permethrin*	+0.30					
(structure: Cl, Cl, COOCH₂, O)		0.05	0.05	+	+	18°
(±) c/t *Esbiol*	−0.56					
(structure: CH₃, CH₃, COO, O)		0.1	0.06	+	+	18°
1-R trans, S 1-R-cis- *Phenothrin*						
(structure: CH₃, CH₃, COOCH₂, O)		0.05	0.1	+	+	
Tetramethrin						
(structure: CH₃, CH₃, COOCH₂—N, O, O)		0.2	0.2	+	+	19°
(±) c/t		$\cong 10^{-9}$ M at the nerve				

Table 100. (continued)

Compound	Polarity	LD$_{50}$ mg/fly	Dose applied mg/fly	Neurophysiological effect		
				Uncoupling of moto-neurons in ganglion	Repetitive backfiring of moto-axon to CNS	Transition Temp. from rep. backfire to single potentials

Kadethrin

		0.07	0.3	+	−	
1 R cis E	−1.09					

Barthrin

		1.0	1.0	−	+	10°
(±) trans						
DDT		0.17	1.0	−	+	13°

Electrophysiological investigations in live flies show, that nerve failure due to the action of pyrethroid occurs only in very few motoneurons. Symptoms and electrophysiological effects at lower temperatures disappear at a sharp higher threshold temperature. The observed electrophysiological effects, depending on the structure of the pyrethroid, are uncoupling of the motoneuron in the ganglion and/or repetitive backfiring of the motoaxon to CNS and flight neuromuscular junction, showing that pyrethroids are nerve ending-poisons [652] (Fig. 18).

Much higher concentrations are required to affect the axons. Table 100 shows the neurophysiological effects observed with different pyrethroids. These effects are not correlated with toxicity, they merely show the advent of toxicant to the target site.

Further analysis of symptoms by a QSAR study of esters

1 Rt

in synergized cockroaches show, that the convulsions are not correlated with neuro*excitation* and lipophilicity of the toxicant, but with neuro*blocking* potency and lipophilicity; and that the lethal effect is neither directly correlated with neuroexcitation and lipophilicity nor with neuroblocking potency and lipophilicity [653].

The visible stages of intoxication, together with the neurotoxic symptoms observed electrophysiologically, are transient lesions, from which the insects can recover.

Death is caused by general impairment of homeostasis or secondary lesions, especially tissue damage followed by loss of water. A sudden threefold increase of diuresis was observed after applying a sublethal dose of 2.8 μg permethrin per locust at the same time as repetitive firing started. Knockdown followed shortly after [654]. The repetitive discharges apparently stimulate the release of a diuretic hormone. Therefore, the neurosecretory system may be the most important target site for the lethal loss of control over metabolism.

6 In-vitro Neurotoxicology/Electrophysiology

Investigation of Neurotoxic Effects on Nerve Preparations

As early as 1942 [655] it was established that pyrethrin is a neurotoxicant in insects [656].

Electrophysiological studies of the effect of pyrethroids on nerve preparations have been shown to be an extremely useful tool in studying the neural disturbances typical for pyrethroids and similarly for DDT, which is related functionally to type I pyrethroids. They can be found not only in nerve preparations from insects, but also very similarly in other invertebrate nerve preparations [657, 691] like the ones from crayfish, leech, [658] squid and others. A concentration of 10^{-10} M pyrethrin I, the natural compound, increases ganglionic activity in the cockroach, 10^{-7} M causes blockage of nerve conduction in giant nerve fibers [659].

This very low concentration of the natural toxin, sufficient to cause in-vitro neural effects, is comparable with the toxic dose for the whole animal in vivo. A typically applied LD_{95} of pyrethrin I gives rise to a 10^{-11} M concentration in the hemolymph of the poisoned insect [660].

However, this concentration is by far not sufficient to have any effect on giant axons of cockroaches.

A typical effect is the repetitive activity on central and peripheral nerve fibers, which leads ultimately to nerve blockage [658, 661].

However, studies in giant fibers from crayfish, leech and cockroach, and comparison of toxicity data in housefly, do not reveal a clear connection of neurophysiological data from nerve preparations and toxicity of a diversity of pyrethroid-structures [662–665, 171].

It was therefore hinted, that giant axons may not be the sites of lethal action [665]. Some toxic pyrethroids have even no action on certain nerve fibers, or have completely different effects on the nerve [664]. Other only slightly toxic pyrethroids have strong effects on nerve action [664].

It could be shown, however, that very small changes in the chemical and sterical structure of pyrethroids do cause dramatic changes in insecticidal and neuro-physiological properties and that the same stereospecificity as for toxicity is needed. Certain aspects of the different neural effects on isolated nerves, such as the recovery of a nerve, are directly correlated to the polarity of the pyrethroid applied [665, 171].

In addition, toxic effects on whole insects, such as speed of action leading to knockdown of the insect, and the lethal effect, are also a function of polarity [666, 667].

Thus, the more polar, but less toxic natural pyrethrin II is a better knockdown agent than the more toxic pyrethrin I.

The more advanced pyrethroids of the metaphenoxybenzylesters like deltamethrin also elicit intense repetitive activity on certain other nerve systems, e.g. the lateral line sense organ in the frog. This effect is due to the prolongation of the sodium current through the membrane [668].

Furthermore, the steady state potassium current [171] and especially the inactivation of the Na-current are also inhibited by a pyrethroid, leading to a delayed repolarisation of the nerve membrane, if any. In locusts there was a clear correlation between polarity, onset of knockdown and onset of neurophysiological effects [171].

Locust nerves are 10 times more sensitive than leech nerves. In the locust, one finds, that an optimum of polarity is needed for the highest action. The most polar pyrethroid was not the most active one in onset of membrane depolarization. Comparative studies on the leech, which is very susceptible to pyrethroids, being knocked down with 0.1 ppm, resulted in a different ranking of active pyrethroid structures. The more polar ones are more effective in inducing depolarizing bursts of action potentials as multiple spikes, and they acted faster, than the less polar compounds. According to a QSAR study [171] these polarity differences explain 60% of the variation in knockdown and nerve action. Comparison of deltamethrin and descyanodeltamethrin in the leech system, which is easy to experiment with, gave some hints of the importance of pharmacokinetic factors. While deltamethrin in-vivo is 15 times more active, the concentration at the nerve in the saline, containing 5 μM pyrethroid, for eliciting the observed effects was the same. Again, no correlation to the lethal effect could be found. Nerve block due to irreversible depolarization was supposed to be the fatal cause.

The known pyrethroids were classified in two groups according to their different effffects produced at nerves in vivo and in vitro [669, 674], whereby type I are more effective on nerve preparation than the more insect-toxic type II pyrethroids (page 128).

Type I cause repetitive activity after a single stimulus of the nerve and a permanent depolarization of the nerve membrane at higher doses.

Type II cause a steady depolarization with smaller amplitude and being constant for much longer times, without repetitive activity. The highly insect-toxic α-cyanopyrethroids belong to this group, while the other ones, like allethrin, tetramethrin, resmethrin and other non-cyano-pyrethroids belong to type I. These two classes also produce also two different groups of toxicological symptoms in insects [670] and in other animals [675, 562] as mentioned before. However, this separation is not clear-cut, there are transitions and also exemptions in certain experiments. In both groups there is the remarkable stereospecificity of action, as found for the insect toxicity.

Analysis of the repetitive firing of type I pyrethroid and DDT showed a good correlation to knockdown properties, but a negative correlation to lethal efficacy [676]. There was a positive correlation of toxicity of type I and type II with increase of miniature excitatory postsynaptic potentials (mepsp), which are signs of nerve terminal depolarization. Toxicity has a negative temperature coefficient, and so does mepsp. Kdr-resistant insects, with a modified binding site as the cause of resistance, are also resistant to mepsp-increasing activity, according to their level of resistance. The fatal stage of a pyrethroid-poisoned insect (*Musca domestica, Culex quinquefascia-*

Table 101. Correlation of miniature end plate potentials (EC_{50}) and lethal effects of some pyrethroids in *Heliothis virescense* larvae

Pyrethroid	Lethal action LD_{50} (ppm)	Neurophysiol. effects EC_{50} mep $(10^{-6}$ ppm)
Cyhalothrin	0.32	5
Deltamethrin	0.35	6.6
Cypermethrin	2.3	28.3
Fenvalerate	3.3	32
Permethrin	10.5	31
Ethofenprox	35	130

tus, Anopheles stephensi, Trichoplusia ni) starts with an increased mepsp rate, followed by neuromuscular block and finally by paralysis. The measurement of mepsps was even proposed as an in-vitro assay for pyrethroid insect toxicity e.g. for *Musca, Heliothis* or *Diabrotica* [677] and was used as such, [678] as examplified in Table 101.

A quantitative analysis of neurophysiological events on the crayfish axon showed, that the enhancement of spontaneous discharges has strong effects on synaptic regions, while induction of repetitive discharges is an effect on the axonal membrane, [679]. These events are dependent on the structure of the substituted benzyl − 1 R trans − chrysanthemates investigated. The most important influence was found to be a position-specific effect of the substituent in the benzyl ring. The onset (t) of neuroexcitation after application (concentration c) of pyrethroids as a measure of the penetration rate was found to obey the following equation:

$$\lg \frac{1}{t} = 0.49 \lg \frac{1}{c} - 0.094\pi + 0.349\sigma - 5.2$$
$$(n = 33, s = 0.19, r = 0.93)$$

This means, lower hydrophobicity ($-\pi$) and electron-withdrawing properties (σ) facilitate penetration [679]. The positive sign of the sigma-term of the latter property points to a bionucleophile as the penetration partner. The concentration for lethal action (LC) of this series toward the cockroach and neuroexcitation at concentration C on the crayfish axon is shown by the equation:

$$\lg \frac{1}{LC} = 1.05 \lg \frac{1}{C} + 0.503\pi - 0.258$$
$$(n = 28, s = 0.439, r = 0.91)$$

In this series of pyrethroids, the variation of the neurotoxic effects and lethal activity could be attributed to differences in hydrophobicity.

A closer analysis of the structural influences of substituted benzyl chrysanthemates on membranes potentials in the giant axon of crayfish revealed three different structure-activity sets:

1. ortho-substituted benzyl esters caused a deceleration of the falling phase of the action potential, which is responsible for repetitive activity, but it is less potent than the next effect.

2. The alcohol moieties

caused an elevation of depolarizing after-potential more lowly, which is not directly linked with repetitive activity.

3. A third group caused a mixture of the both effects.

The kinetics of these effects are governed only by penetration through a number of barriers, which is facilitated by lower hydrophobicity. Other structural features play a minor role. The cause of these effects is, that there are two distinct binding sites. In this study [680], α-cyano-esters were not taken into account.

Repetitive firing occurs with 10^{-8} M allethrin in only 10% of the nerve fibers of cockroaches [681].

The LD_{50} and the multiplicity of nerve-signal spiking in the labella hair of the fly *Lucilia cuprina* in the class of DDT-pyrethroid-hybrids is somehow correlated [194].

More recent results [682] show, that in locust and leech there is no principal difference in the action of pyrethroids of both groups (allethrin, deltamethrin, descyanodeltamethrin) on the neurons. All the same responses were elicited, depending on physicochemical properties, dose, time of measurement after application of the pyrethroid and the type of nerve preparation. A general basic effect on a specific binding site within the nerve membrane, common for all pyrethroids, may be the primary cause for secondary changes in conductance of different ions, including Na.

The type II pyrethroid deltamethrin applied at a higher concentration of 10^{-7} M was found to produce repetitive firing of the nerve, typical for type I pyrethroids not only in frog [668], but also in the ventral nerve cord of the central nervous system in cockroaches [683]. Apparently in-vitro nerve preparations are rather less sensitive to pyrethroids than in vivo.

The fundamentally similar action of pyrethroids on nerves of arthropods results from the significant correlation of the neurotoxis of a series of pyrethroids on crayfish and knockdown activity to houseflies [685].

The discrepancies between toxicity and performance in a given nerve preparation are sometimes impressive. While the topical LC_{50} of deltamethrin against cockroach is about 5000 times lower than for 1-R trans-tetramethrin, the former is 10^6 times less effective in repetitive firing in the sensory cercal nerve. There are two explanations, either the much more toxic compound acts directly on sensory receptors, or it penetrates much faster and more efficiently through the cuticula.

In fact, the poisoning symptoms after topical applications are clearly related to the decrease in electrical activity in the central nervous system [686].

The problem of the correlation of electro-neurophysiological response and toxicity was shown also in the comparison of different effects of the very similar compounds, the ($\pm$) transchrysanthemic ester of pentafluorobenzyl alcohol and the corresponding permethric ester (racemic fenfluthrin) [687]. Neurotoxicity, knockdown potency and mortalities were found to be in the ratios of 1:54, 1:2.6 and 1:2.4 respectively. While the former compound produced dose related heavy repetitive activity (threshold concentration 10^{-10} M) and large bursts of action potential in an increasing number of spikes, the latter (threshold concentration 10^{-11} M) did not show the number of spikes to be dose-dependent. Most important for toxicity here is the threshold concentration at which the impairment of nervous balance starts. The neurophysiological response does not seem to be so important for lethal action.

Although impressive effects of pyrethroids are seen on axonal nerve preparations, the micromolar concentrations needed to elicit them are higher by several orders of magnitude compared to the actual toxic dose against cockroaches applied in practice. Furthermore, the sensitivity of the sensory cells of cercal nerves in the cockroach is much higher than for axons. The threshold concentration for repetitive firing is again higher than the toxic dose. Thus, the increased sensory stimulation by pyrethroids does not explain the toxicity [688].

One problem in comparing neurotoxic effects on insect nerves in-vivo and in-vitro is the uncertainty of the actual dose applied to the nerve. By applying precise doses of pyrethroids by infusion into living locusts, it could be shown, that the electrophysiological response in-vivo and on prepared ganglia, as well as the neurotoxicological and toxicological symptoms, are indeed closely correlated [684].

Other reports [689, 673] on in-vivo results on cuticular mechanoreceptors of the cockroach showed that they are affected by a 150–5000 times lower concentration than the LC_{50}. According to the response after mechanical stimulation of intoxicated cockroach a clear distinction of type I and type II pyrethroids was evident, as shown in the following Figure 19:

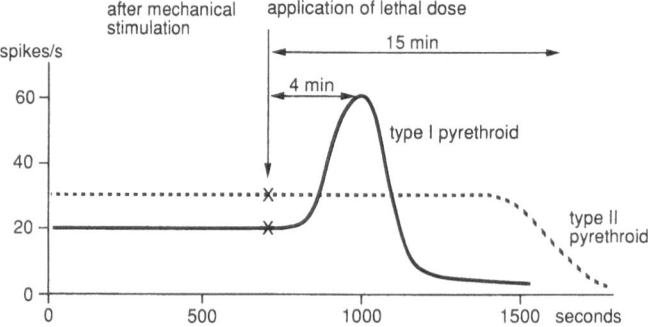

Fig. 19. Different intensity of nerve signal spiking as electrophysiological respons of the application of type I and type II (α-cyano) pyrethroids

While this book went to print a summary of the different electrophysiological activities of type I and type II pyrethroids in different parts of the insect nerve system was presented by Narahashi, as shown in Figure 20 [823].

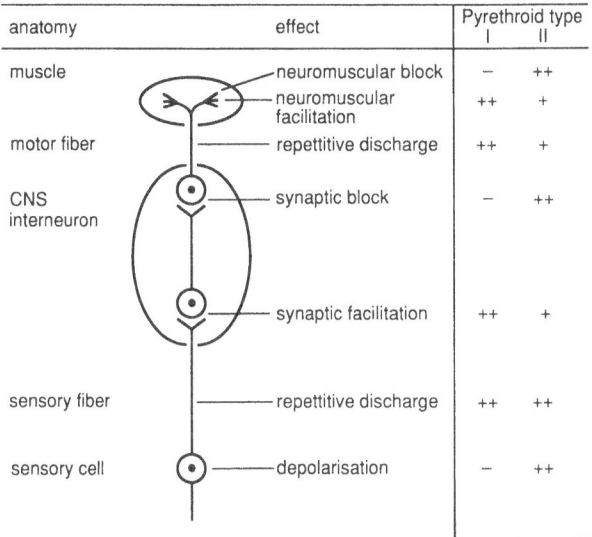

anatomy	effect	Pyrethroid type	
		I	II
muscle	neuromuscular block	–	++
	neuromuscular facilitation	++	+
motor fiber	repettitive discharge	++	+
CNS interneuron	synaptic block	–	++
	synaptic facilitation	++	+
sensory fiber	repettitive discharge	++	++
sensory cell	depolarisation	–	++

Fig. 20. Sites of action of type I and type II pyrethroids in the nerve system of insects

7 Molecular Neurotoxicology

7.1 Pyrethroids as Ion Channel Effectors

7.1.1 The Effect of Pyrethroids on Sodium Mediated Nerve Conduction

From the beginning of the investigations of the mode of action of pyrethroids, most of the evidence led to the explanation of the involvement of sodium channels and the sodium ion currents diffusing through them [690, 691, 692, 668]. The electro-physiological tools, the easy access to nerve fibers in-vitro and live animals, in connection with the increasing number of interesting structurally different pyrethroids, yielded a complex field of data, sometimes conflicting, mostly corroborating. At the moment the evidence of the sodium channel as the lethal target site of pyrethroids in insects and other animals is supported by the majority of investigators.

A whole wealth of data on the action of pyrethroids was first presented at the conference Insect-Neurotox 79 in York, an almost historical event for the interdisciplinary research on molecular neurobiology by chemists, entomologists and neurophysiologists.

Since so many results on the effects of pyrethroids on Na-channels at different concentrations, on different biological subjects and from different experiments have been produced within the last ten years, it is worth devoting a chapter to this topic, which may be important for other neuropharmacological problems in the future.

Normal nerve cells have excitable membranes, caused by a delicate balance of different ion concentrations on both sides of the insulating lipid membrane, leading to a resting potential of 40–60 mV in insects.

The exchange of information from the outer environment and the information processing nervous system, and within the central nervous system is mediated by membrane bound ion channels. The nerve conduction is the propagation of a transiently changing potential running along the nerve fiber by the sudden exchange of Na-ions through the membrane from the high concentration in the extracellular medium to the very low Na concentration in the cytoplasma channeled through the sodium channel, a membrane bound protein. A channel opens under the influence of a giant electrical field strength, caused by the opening of neighbouring Na-channels.

The resting potential of nerve membrane is composed of the Na and K-gradients, produced by the energetically neutral diffusion of K through the membrane, and by constant pumping, using up energy, [693] as is necessary for the Na gradient.

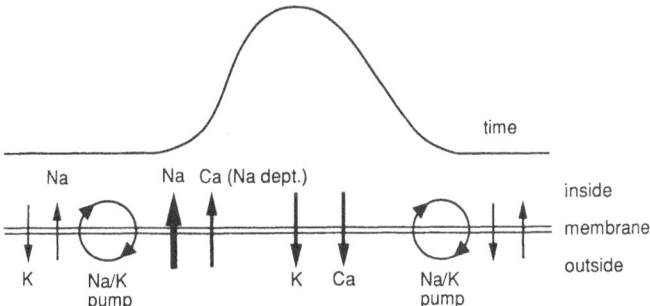

Fig. 21. Consecutive transmembrane ion currents during the formation of an nerve action potential

The action potential involves a complex interplay of mutually delayed ion currents through the membrane, as depicted for the cations in Fig. 21 [694].

There are many different ion channels and states of channels in a nerve cell. Voltage gated Na channels in higher animals are found mainly in the nodes of Ranvier and satellite cells surrounding the axon (glia cells in CNS, Schwann cells in the peripheral nervous system). Glia cells are presumably the site of synthesis of Na-Channels and K-Channels as well [695].

Voltage gated K-Channels are found in the axolemma, together with very few Na-Channels. Figure 22 shows this situation.

In one node of Ranvier in amphibia, about 1000 to 1500 Na-channels/μm^2 are concentrated. Insect nerves, however, having no nodes, show an average density of 100 Na-channels/μm^2, distributed over the whole surface of the nerve cell, which is involved in signal transmission [696]. This difference, together with the much better insulation of the axon by the surrounding satellite cells, makes nerve conduction in higher animals much more effective than in insects.

The area of one node is 30 to 60 μm^2, containing therefore about 70000 channels in amphibia, about 20000 in rats [627]. Another guess leads to a mean distance between the Na-Channels in nodal areas of about 1000 Å [696].

If only 1% of the channels need to be modified, to give rise to a grave disturbances of nerve conduction [692], this would result in a mean distance of two channels, modified by one pyrethroid molecule of about 10000 Å. To inhibit the proper relay function of one node 200 to 700 pyrethroid molecules would be sufficient.

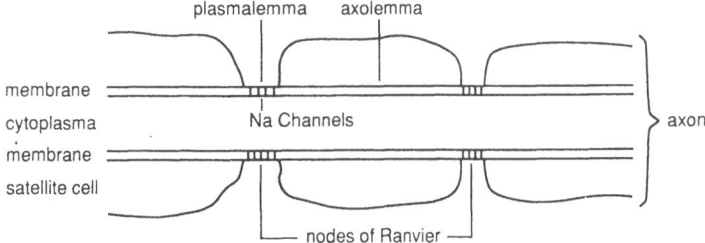

Fig. 22. Nodes of Ranvier in a nerve fiber of a vertebrate as sites of sodium channels

There are reports of almost incredibly low concentration of pyrethroids affecting certain in-vitro nerve preparations, e.g. in the cercal nerve of the cockroach, where methanotetramethrin caused measurable effects at 10^{-18} M [698], and from corn rootworm larvae [699], where the following effective concentrations were found: permethrin $8 \cdot 10^{-16}$ M, cypermethrin $5 \cdot 10^{-17}$ M, cyhalothrin 10^{-18} M. This extremely low concentration corresponds to about 100 molecules/ml, or in macroscopic terms, about 50 mg/km^3. Much higher concentrations were necessary for similar effects in *Heliothis* larvae. Neurobioassay of *Diabrotica* larvae showed their extreme sensitivity to pyrethroids, where 10^{-16} to 10^{-18} M concentrations caused miniature endplate potentials [677] in contrast to the far less sensitive *Musca domestica* or *Heliothis virescens*. Similarly, isolated leech nerves are affected by 10^{-15} M deltamethrin and 10^{-12} M Permethrin [677]. However, extreme and particularly critical care has to be taken in analyzing highly diluted aqueous solutions of pyrethroids in order to coroborate such homeophatic doses (see page 119). (Limits of detection by GC/ECD in 1986 for cypermethrin for example was 10 ppb in animal tissue [821, 836]. In comparison 10^{-10} M permethrin, having similar physical properties as cyhalothrin correspond to about 0.04 ppb). The dilution techniques have also to be, scrutinized.

The insect nerve preparation needs to extract in to the lipophilic membrane only one ml of the saline, containing such a low concentration of highly lipophilic pyrethroid, and redistributes it to the very specific binding site, in order to block 1% of the Na-channels present, disturbing the whole nerve conduction. The determination of the actual concentration of labeled deltamethrin in the nerve tissue after application of $2 \times LD_{50}$ to a *Spodoptera* larva yielded the value of about 0.07 μm after 12 hours [844]. However, the toxicological relevance of this nerve effect is sometimes questioned due to the observation that cis-tetramethrin, cyclopropanated on the olefinic bond of the chrysanthemic acid moiety is less toxic than the parent compound, but neuroactive at 10^{-18} M [700].

7.1.2 The Action of Pyrethroids on the Presynaptic Sodium Channel

Each channel is made of a transmembrane protein, enclosing a 5 Å diameter water-filled pore in the open state, just wide enough, to let small cations like Na^+, Li^+, NH_4^+ and formamidinium pass through.

According to a recent proposal deduced from the Na-channel-mRNA from rat brain, this protein is a single strand, made of about 2000 amino acids, whose rolled up topology and charge distribution is shown in Fig. 23 [701, 822].

An interesting feature is the 4-fold occurrence of highly positively charged lipophilic transmembrane segments. They can easily move like a spring towards the cytoplasma at the beginning of depolarization of the membrane as a voltage sensor, carrying thereby 4 to 6 positive charges across the membrane and making room for a pore or just opening the passage. This corresponds to the observable gating current needed to open up the channel quickly for the sodium traverse. In other words, this is the activation of the channel by opening the fast m-gate. Since the inactivating h-gate closes slowly, there is time enough (some milliseconds) to allow the almost unhindered passage of about 13000 Na-ions/msec [697].

The old "molecular spring" model according to Weiss should be taken into account once more [701a].

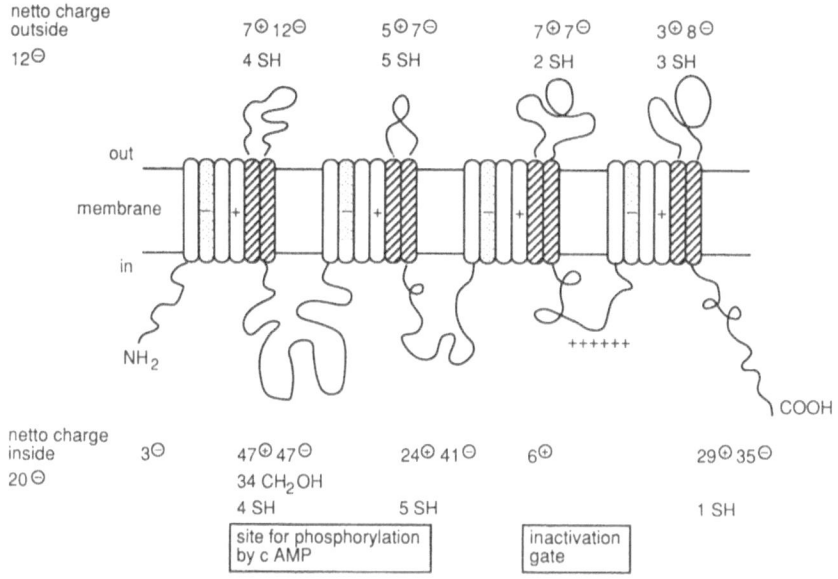

Fig. 23. Rolled up topology of proposed structure of the sodium channel protein in the nerve membrane of the rat. Hatched areas are contacting the surrounding lipids. The actual ion channel is formed by the dotted bars.

This sudden permeability change of the nerve membrane is independent of the generation of metabolic energy in the axon, since metabolic inhibitors, like cyanide or azide ions, showed no effect [692]. Chemical activation by phosphorylation seems to be likely. In the above mentioned protein model there is one cytoplasmic loop very rich in serine and carboxylates, best suited as a substrate for the phosphorylating agent ATP.

In addition, insect Na-channels are Ca sensitive [702]; this sensitivity diminishes to a large extent in resistant strains.

Pyrethroids, e.g. allethrin, keep Na-channels open [716a], in the manner the snake venom condylactis toxin does, as deduced from the electrophysiological experiments on isolated nerve fibers. In the crayfish axon they prolong the Na-current during excitation beyond the termination of the outward repolarizing K^+-current, which leads to the observed repetitive discharges of polarized nerves [668, 692], resulting in more or less long trains of spikes on the recording screen. In addition they also modify the channel-opening mechanism [703, 686].

At higher concentrations, both the Na- und K-currents are suppressed. The effect of allethrin in frog myelinated peripheral nerve membrane preparations is seen within a few minutes after application. The number of Na-channels affected is proportional to the number of open channels, where both, the fast m-gate and the slow h-gate are open. Allethrin affects only the closing of the m-gate and keeps it open for a long time [668], allowing a larger portion of Na-ions to travers the membrane, thus depolarizing the distribution of charges on both sides of the insulating membrane. The normal functioning membrane potentials oscillates with the discharges between

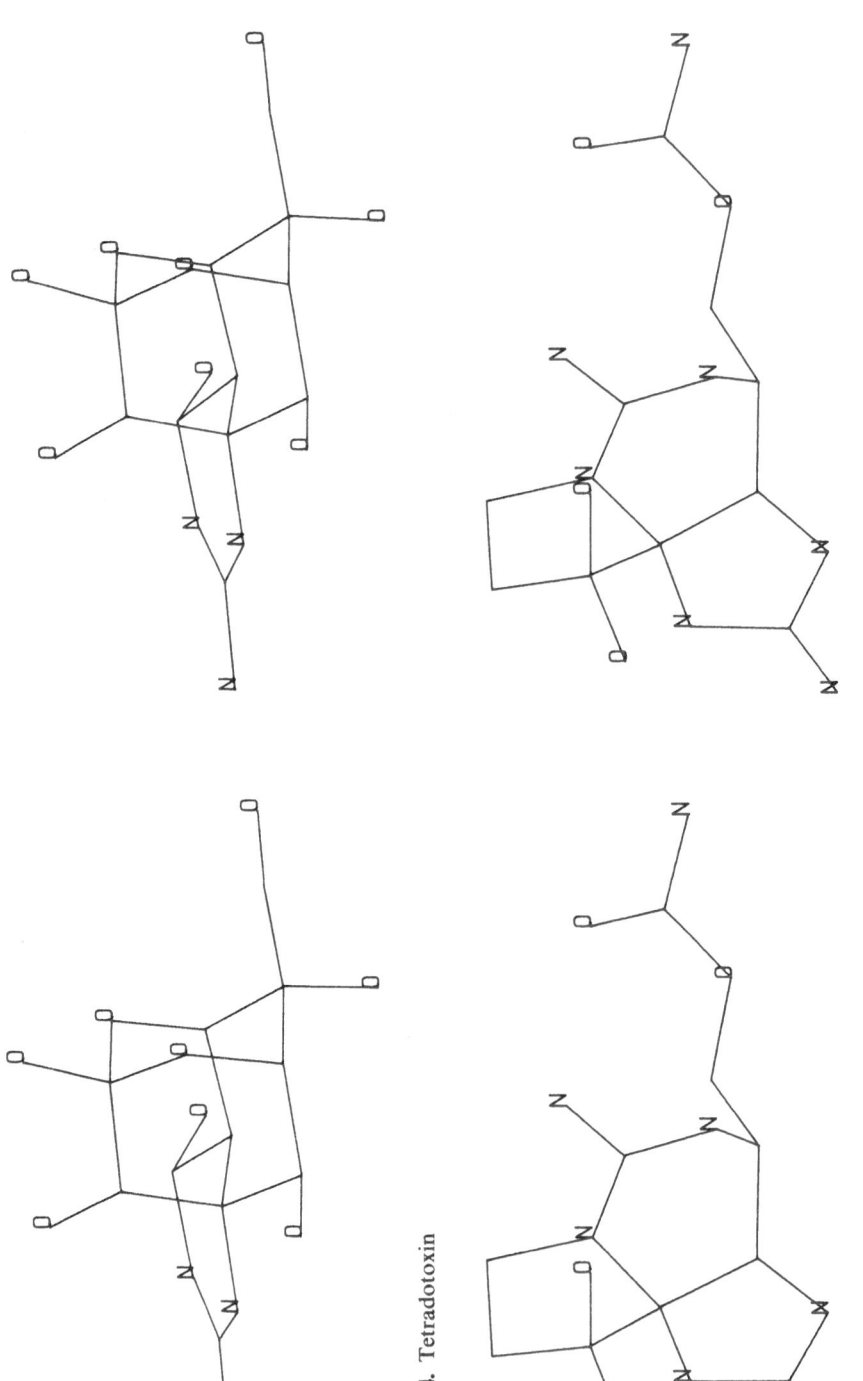

Fig. 24. Tetradotoxin

Fig. 25. Saxitoxin

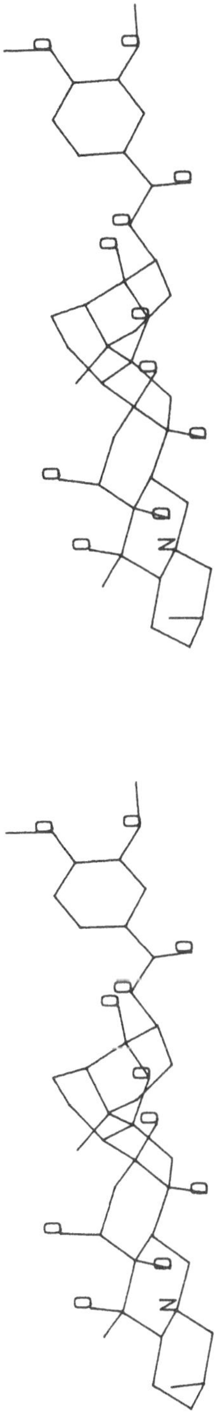

Fig. 26. Veratridine

Fig. 27. Grayanotoxin

Fig. 28. Batrachotoxin

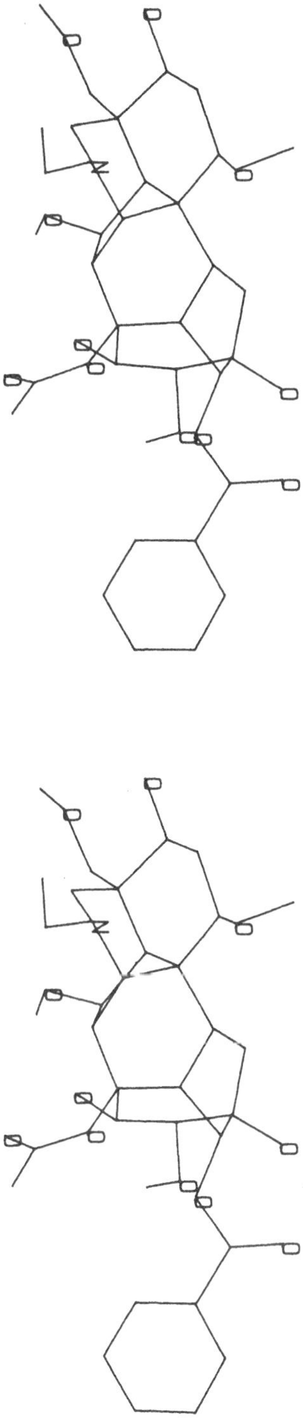

Fig. 29. Aconitin

−80 mV and −10 mV. The ion selectivity and energy profile of the Na-channel itself
are unaltered by the binding of a pyrethroid molecule [704].

The specific Na-channel blockers tetrodotoxin and saxitoxin [750], acting at the
extracellular face of the channel peptide, also block at 10^{-7} M the pyrethroid effect
on the nerve in the cockroach [706, 707], but not the effects of DDT.

The first effect of pemethrin is an intermittent intense inward Na-current at those
parts of the nerve membrane, which are not specialized for propagating action
potentials and which are located at central interneurons, thus influencing other
post- and presynaptic events [708]. Since insects have only few neurons involved in
central nervous integration, the deregulation of only a very few will have grave
consequences for the whole balance of the nervous system.

This channel protein and its very close vicinity, the firmly attached belt of lipids
(distance protein-lipid about 3.5 Å), is the target site of a number of interesting and
diverse toxins, mostly very lipophilic and rigid polycyclic compounds with no or
limited freedom of molecular flexibility. Present knowledge stems from electrophysiological investigations. The term molecular wedge, originally coined for DDT
by Holan, may describe the function of most of these compounds.

Several distinct functional target sites have been identified so far, where different
compounds bind and cause different effects on the sodium current.

The complex stereo-formulae of these polycyclic toxins are drawn by computer,
to avoid the awkward projections usually seen. The hydrogen atoms are not shown.

Target site I is situated at the extracellular outer side of the sodium channel.
Compounds acting here, like tetrodotoxin (irreversible) (Fig. 24), and saxitoxin (reversible) (fig. 25), block the sodium current flowing through the open channel.

Target site II seems to be situated at the channel-proteinlipid interface. Certainly
several binding sites are involved for compounds, which may act as competitive
or non-competitive effectors. They open up closed channels or keep them open,
thus they prolong the sodium current by different mechanisms and may cause
repetitive firing of the nerve.

Molecules acting at these binding sites are structurally very divers, although
sometimes very similar in their electrophysiological action. Veratridine (Fig. 27)
and DDT and some pyrethroids as well are typical examples. Veratridine activates
(= opens) maximally 8% of all sodium channels in neuroblastoma cells [710] and
causes repetitive firing.

Pyrethroids [710, 712, 713] open up channels and/or keep them open and cause
repitive firing, like DDT [712, 713]. Some pyrethroids like bioresmethrin, cismethrin
or biopermethrin antagonize the action of the pyrethroid kadethrin on the sodium
channel in neuroblastoma cells, opened up by veratridim [710].

Other site II channel openers are grayanotoxin (Fig. 27), batrachotoxin (Fig. 28),
aconitine (Fig. 29) and unsaturated isobutylamides Fig. 30. The latter, having only
moderate insecticidal activity, antagonize veratridine induced transmitter release
in synaptosomes and cause repetitive firing in axons, which can be blocked by
TTX. They are more toxic to super-kdr- insects and more active on sodium channels
from super-kdr-flies as well [714, 715, 716, 353].

Condylactis toxin [716a] and procaine also have their target site here.

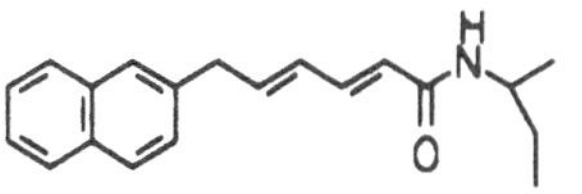

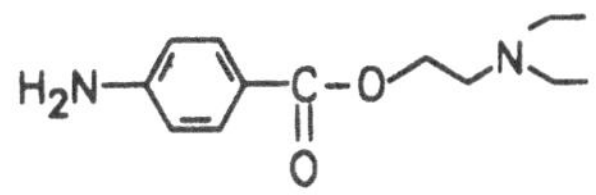

Fig. 30. Unsaturated isobutylamide (left) and procaine (right)

Target site III is affected by peptide toxins from scorpions and sea annemone. They bind at the cytoplasmic site of Na-channel, slow down Na inactivation and enhance persistent activation by veratridine.

Target site IV interacts with the red tide toxins [711], the brevetoxins (Fig. 31 and 32). Brevetoxin A is extremely toxic to fish (4 ppb), causes repetitive firing, increases frequency of action potentials and enhances veratridine action. The recently elucidated structure [718] shows a strange and rigid molecular rod like shape. Extending for about 30 Å in one dimension it could span a good deal of the whole membran. Its slimness in connection with its twisted stiff shape on the other side makes it a good condidate for intercalation in transmembran helices, thus blocking the flexibility of the gating machinery.

Crayfish are very interesting objects for investigating pyrethroids. They are very sensitive to DDT, and, for example 1 R-trans-tetramethrin injected at 0.02 mg/kg. They are 100 to 800 times more sensitive towards pyrethroids than the housefly, 1800 times more than the locust with topical application. Investigations showed, that all macro- and microscopic [692a] effects of pyrethroid poisoning are to be found in all the species mentioned. The visible signs are hyperexcitation, ataxia, convulsions and paralysis. Here, on whole nerve cords, single nerve fibers and neuromuscular junctions, respectively, it could be shown, that the repetitive discharges triggered by the depolarizing afterpotentials caused by a threshold concentration of $3 \cdot 10^{-9}$ M tetramethrin, originate from presynaptic nerve terminals. This causes prolonged after discharges in presynaptic nerve cells, lasting sometimes for more than one hour. Within a nerve bundle the single fibers may behave very differently to pyrethroids. Some neighbouring fibers are even completely unaffected. Within one sensitive fiber, only a fraction of the Na-channel needs to be modified by binding the toxicant, in order to explain the overall result seen in the nerve. Only a difference of a few mV is necessary to produce the repetitive firing. This is already accomplished, when 1% of all Na-channels are modified using a threshold concentration of 10^{-9} M trans-permethrin for example.

The kinetic analysis of 10^{-5} M tetramethrin on squid axons revealed, that 5 different states of Na-channels are involved simultaneously, namely resting state, open state and inactivated state, and their pyrethroid-modified counterparts, which transform into each other with a slow or fast time constant. Thus pyrethroids bind at closed and open channels [719].

This transformation depends on channel integrity, since partial enzymatic proteolysis abolishes the slow inactivation process, the closing of the channel. It greatly enhances the formation of the pyrethroid modified open state.

The stereospecific effects on Na-conductance, caused by very active pyrethroids, were first classified in two groups, as previously mentioned: [669–674, 720, 721].

Type I: non α-cyanopyrethroids
Type II: α-cyanopyrethroids

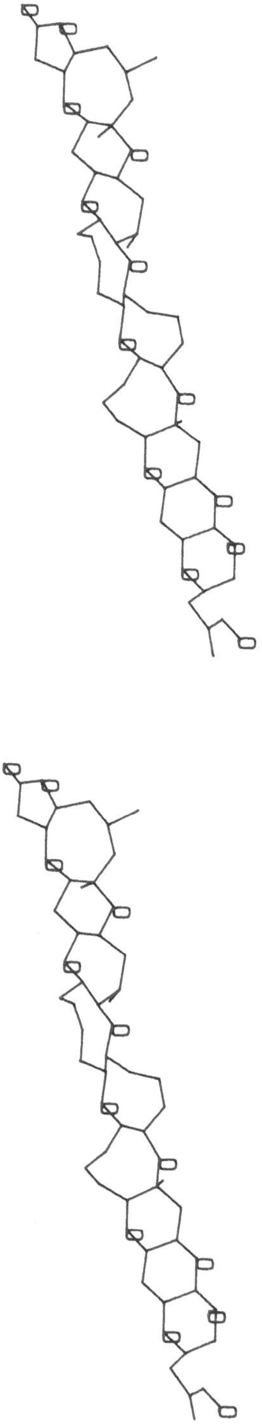

Fig. 31. Brevetoxin A

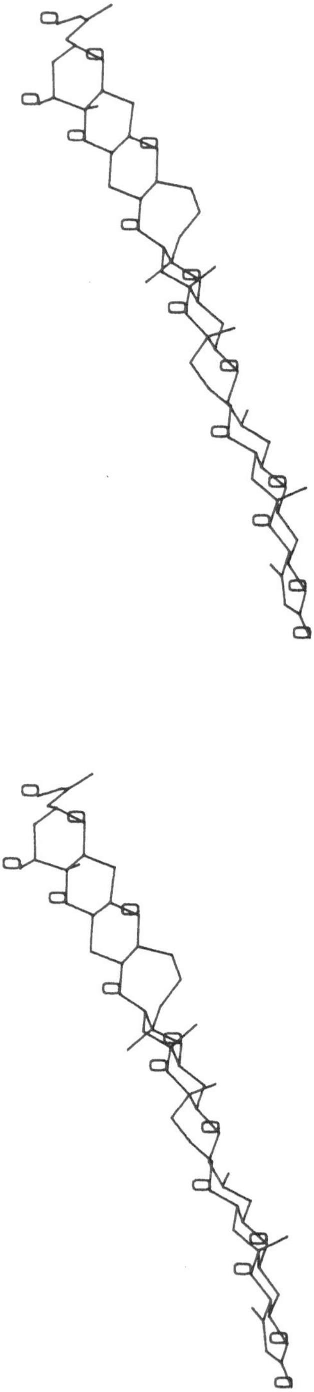

Fig. 32. Brevetoxin B

Table 102. Opening time of pyrethroid-modified sodium channels of different origins

a)	From frog nerve [720, 721]	Opening time
	DDT	3.7 msec
	1 R trans Permethrin	0.4
	± Allethrin	9.8
	1 R Cismethrin	20.7
	1 R cis Permethrin	28.7
	1 R cis Fenfluthrin	100
	Cyphenothrin	290
	S, α S Fenvalerate	602
	1 R trans α S Cypermethrin	1020
	1 R cis α S Deltamethrin	1450
	R α R Fenvalerate	no effect
b)	From axon [674]	
	DDT	9.5 msec
	Tetramethrin	620
	Phenothrin	1340
	Cypermethrin	several minutes
c)	From neuroblastoma cells [674] at (5 °C)	
	untreated	3 msec
	Tetramethrin	10 msec

Since then, other pyrethroids have been synthesized, which combine both properties and have intermediate pharmacological qualities.

Type II, the most toxic insecticides, cause very long trains of efferent nerve pulses in peripheral frog nerves induced by a very slow decay of the Na after-current, which is negatively correlated to temperature. These trains may last for many seconds. The ranking of the time constant (opening time of channel) of the Na-tail current in this nerve preparation [720, 721, 722] using gigaohm seal patch clamp techniques is interesting (Table 102).

These effects are more pronounced in sensory fibers than in motor fibers. Repetitive discharges are the primary cause of hyperexcitation at the synaptic and neuromuscular junction [674]. The threshold concentrations to elicit them depend on the structure of the pyrethroid. However, they bind in an all or none fashion, i.e. the channel is either completely normal or completely modified with different time constants reversal of the modification.

The variation in the time constants of the Na current the variation in poisoning symptoms in mammals and insects.

In the squid axon the effects caused only by the insecticidally active isomers of tetramethrin are fully developed in a steady state after 15 min [723]. They are slowly, but almost completely reversible.

According to dose-response curves of the crayfish giant axon with tetramethrin isomers more than one binding site is involved. The inactive(−) S-cis-isomer also binds as a competitive antagonist to the active R-cis-isomer, without changing the function of the channel. Furthermore, it seems that the two active isomers 1-R-cis and 1-R-trans tetramethrin bind as agonists at different sites of the Na-channel. In addition, a third negative allosteric site to account for the non competitive antagonism of the inactive 1 S-isomere is apparently involved [724].

In the giant axons of crayfish, the type II pyrethroid leave the Na-channel in the open state without causing repetitive activity [723 q].

They depolarize the membrane until blocking of action potential occurs. Type I pyrethroids modify the channel transiently, causing repetitive nerve activity.

Within the group of type II pyrethroids, very different effects are produced in certain frog nerve preparations: deltamethrin increases stationary Na conductance, cypermethrin blocks it, and fenvalerate has still different effects on the threshold level of nerve excitability [721 a].

The 3-Phenoxybenzylesters without a cyano group and the pentafluorbenzyl ester fenfluthrin [722] have an intermediate position in their effects on the axon, thus showing a continuous transition from type I to type II structures.

The occurrence of repetitive activity varies markedly within the peripheral nervous system. In many fibers from different parts of the nervous system of the frog or mouse neuroblastoma cells no difference between α-cyano and non-α-cyano could be found [722]. Also in single channel recordings of opening time [674] the continuous change of properties from DDT to deltamethrin could be shown. It may therefore not be necessary to invoke two different modes of action on Na-channels to explain the striking differences of the observed electrophysiological and toxociological effects.

Investigations of the binding properties of nerve cell cultures from different nerve tissues at high concentrations of pyrethroids, using 10^{-5} M labeled deltamethrin showed the following order of affinity:
dorsal root ganglion > peripher nerve fibers, spinal cord ≫ muscle. Satellite cells always have a higher affinity to pyrethroids than neurons. The same binding was observed in the myelin sheath, axoplasm, nodal or paranodal regions [722].

The duration of trains of repetitive discharges in frog nerves is limited by Na-repolarization, i.e. the accumulation of Na ions on the outer side of the membrane. A slow accumulation results in long trains of discharges [722].

Similar investigations of cockroach giant axons again showed that a micromolar concentration of deltamethrin only affects a small number of Na-channels and keeps them in an open state for a long period. However, the non-α-cyano-pyrethroids allethrin, tetramethrin, cismethrin, kadethrin and permethrin induce only brief alterations in a large number of open state channels [705] and prolonged depolarizing repetitive afterpotentials. It seems that the action of pyrethroids on insect giant axonal Na-channels only partly contributes to the overall toxicity. Moreover, sensory axons are much more sensitive. They are affected with picomolar concentrations of deltamethrin [725]. A breakthrough seems likely to come from much more precise investigations.

Investigations of single channels by the patch-clamp technique revealed that deltamethrin and tetramethrin prolong the mean opening time of a channel 8 to 10 fold [726]. The following repetitive discharge can be associated with the poisoning symptoms in insects. The α-cyano-pyrethroid opens the affected channel dramatically longer, leading to complete membrane depolarization. All these events should now explain toxicity [712].

In spite of the extreme effects on the nerve caused by minute amounts of highly toxic pyrethroids, especially in the periphery, this is drastically changed in *resistant* insects. The threshold concentration of permethrin for interference with peripheral nerve conduction is 10^{-12} M for susceptible and higher than 10^{-7} M for resistant fly larvae [706]. The kdr-factor, the reason for this diminished binding, may result from changes in lipid solubility of the nerve membrane or an altered binding architecture at the protein-lipid interface or within the channel protein.

Recent in-vivo electrophysiological investigations of the neural effect on mechanoreceptors of the pyrethroid-poisoned free-walking cockroach, which are directly connected to the CNS without synaptic linkages, actually confirmed the evidence for the Na-channel as the site of toxic action [727].

Bioallethrin effects: − substantial increase in the number of action potentials by mechanical stimulation
                           − prolonged repetitive activity after mechanical stimulation
                           − inhibition of excitability at the site of mechanical stimulation

Deltamethrin effects: − inhibition of excitability of the cellmembrane at the site of mechanical stimulation
                           − transient spontaneous electrical outbursts.

To summarize, most of the evidence at the moment favours the explanation of a primary action of pyrethroids on fast voltage gated Na-channels in certain parts of the insect nerve system, which cause longer lasting secondary, finally lethal effects if the membrane is depolarised for a long period.

Nevertheless, strong effects of pyrethroids on nerve preparations having almost no Na-channels showed, that other cation channels (K, Ca, slow Na-channels) may also be affected in a very similar manner [728, 729].

Even in TTX blocked preparation a pyrethroid may still have strong effects [730].

Other additional, even more sensitive target sites cannot be completely excluded, not to speak of the very different sensitivity of Na-channel -subpopulations somewhere in the nervous system.

7.1.3 Neurophysiological Alterations in Pyrethroid Resistant Insects

As soon as the photostable pyrethroids came into use for commercial application in agriculture and public health, they had to face more or less resistent insects in certain areas, where they had never been in use before. It soon, turned out, that this resistance was mostly caused by the preceding intensive use of DDT. It was characterized by a limited *knock down rate* of pyrethroids. (kdr-resistance or even super kdr). This resistance is the most important resistance-factor in houseflies.

Neurophysiological investigations of the background to this situation, which is embarrassing for farmers, revealed, that in kdr- or super-kdr-flies the penetration or detoxification of labeled permethrin was unaltered. But the sensory nerve fibers of kdr-fly larvae were 1000 fold, the super-kdr 10000 fold less sensitive in regard to repetitive discharges, intermittent bursting and nerve blocking. The effective concentration was 10^{-6} M, in susceptible larvae 10^{-10} M [731].

The resistance factor was also reflected in the time, in which the flies recovered from knock down: susceptible flies remained in this state for hours, kdr flies were up again within 6 to 12 min., super-kdr flies were up again after a few minutes, a time which is not sufficient for metabolic degradation to a relevant extent.

Another highly resistant housefly strain from the USA showed decreased cuticular penetration for trans − permethrin in addition to decreased nerve sensitivity to several pyrethroids [732].

Membrane bound enzymes, like acetylcholine esterase, normally show a discontinuity in their performance at a sharp temperature step, as a result of a liquid phase transition in the membrane, to which the enzyme is anchored.

In kdr-flies this transition temperature is shifted from 14° to 19°, in super-kdr flies to 21 °C [733].

This property is lost in hybrids between super-kdr and susceptible flies, showing the recessive behaviour of this resistance. The kdr-resistance apparently selects flies with more viscous lipid membranes, maybe by shifting to larger ammounts of the less unsaturated fatty acids in comparison to the normal constituents.

A direct effect of micromolar permethrin or cypermethrin on lipid membranes from spleenic lymphocyte membranes is a decreased packing order of membrane lipids [735] thus lowering the phase transition temperature [736].

In search of the more precise cause for the kdr-factor, the action of different typical Na-channel effectors were investigated.

The nervous system of kdr-resistant fly larvae was 10^{-4} to 10^{-5} times less sensitive to permethrin, veratridine, deltamethrin, and aconitin than that of susceptible ones [743]. None of the more polar, powerful Na-channel effectors (tetrodotoxin or scorpiontoxin), or procaine show this effect [706]. This shows, that kdr-resistance arrests poisoning with Na-channel-toxins at two distinct sites, since these two compounds act at different loci on the channel [737] protein.

Nerve membranes from the heads of susceptible flies containing receptors for DDT and permethrin, bind much more cispermethrin than preparations from kdr-flies, thus indicating, that a reduced number of receptors may be the cause of kdr-resistance [738]. However, kdr-resistant flies do not show a reduced number of Na channels [826].

Studies in other resistant insects also supported an altered target site as the main cause for kdr-pyrethroid resistance.

In a permethrin resistant *Spodoptera* larva (resistance-factor 4) from Egypt, which was not resistant to cypermethrin, no difference in penetration or metabolism was found [739]. This resistance disappears at 10 °C and becomes more pronounced at 30 °C (resistance-factor 12) [740]. Less pronounced was this temperatur effect in resistant *Heliothis* [314a], resistant to fenvalerate, flucythrinate and permethrin. Nerves of resistant *Spodoptera* larvae were more sluggish in developing repetitive discharges at 28 °C. This could be speeded up by less Ca in the saline. Both preparations from susceptible and resistant larvae did not produce repetitive discharges below 19 °C. Another 400 fold resistant *Spodoptera exigua* had two components of resistance. The minor one is synergisable only by oxidase inhibitors and is not changed with temperature. The other one is not affected by synergists, having a negative temperature-coefficient.

The temperatur sensitive *Drosophila* strain *napts* is not affected by pyrethroids like fenfluthrin, MTI 500 or cypermethrin [826].

In cockroaches also, a strain was found kdr-resistant to permethrin, allethrin, pyrethrin and fenvalerate, but not to deltamethrin and cypermethrin [221].

From physicochemical calculations of binding enthalpies of cis-cypermethrin in resistant and susceptible cattle tick larvae it was concluded, that the 6 fold resistance is not related to altered binding, rather to altered access to the site of action [741].

In the future, a search for the cause of the observed resistance effects will bring about more important explantions for the actual underlying causes and the molecular genetic mechanisms.

7.2 Specific Action of Pyrethroids on Other Important Target-sites in the Nervous System

In view of the sometimes conflicting data from electrophysiological investigations of pyrethroid action on nerves, mostly in support of Na-channels being the primary target, it was tempting nevertheless to look for perhaps clearer and alternative explanations for the lethal action on insects. Independent biochemical evidence would be very helpful. Despite the minute amount of Na-channel protein present, the existence of a specific binding site could be shown by a modern receptor technique [738, 742], when radiolabeled pyrethroids were available. DDT and cis-permethrin compete for the same specific Ca-sensitive nerve-membrane receptor site [738, 742] in housefly-, *Chrysopa carnea*- and *Heliothis*-CNS, but not exactly for the same binding site, thus supporting the electrophysiological evidence [743]. The reversibility of this binding depends on the species tested. This binding in the almost pyrethroid-in-sensitive predator insect *Chrysopa carnea* is reversible. It is irreversible in the highly susceptible *Heliothis* and is moreover negatively correlated to temperature. The apparent dissociation constant is $1.2 \cdot 10^{-8}$ M. Chlodimeform increases this specific binding [742].

However, the rank of binding (DDT > cypermethrin > cis-permethrin) differs substantially from the order of toxicity toward these insects.

In mouse-brain-membranes a receptor could also be found which binds del-tamethrin stereospecifically, showing a halfsaturation at 10^{-8} M [744]

A first approach to investigate the molecular aspects of binding at the peptide was undertaken by designing and synthesizing a 24-mer lipophilic polypeptide, which binds DDT rather specifically in comparison to other peptides. For further analysis this peptide-DDT-complex could be crystallized [745], and very recently as the first artificial protein synthesized by *E. coli* by genetic engineering [746].

The near future will certainly bring about a dramatic expansion of knowledge on the function and distribution of ion-channels by combination of single channel electrophysiology and receptorpharmacology on channel proteins from different species, the place within the nervous system and most interesting on artificially altered channel proteins, synthesized via genetic engineering. Many effects found during the work with pyrethroids will be taken up again for a thorough investigation. In the future closer cooperation with chemists and their critical knowledge should make this work more valuable. Some projects in the past, started by very interesting questions, investigated with very good methods, were almost spoiled by the poor sampling of compounds, or by not regarding the stereochemical aspects, thus yielding only limited information.

7.2.1 Effect of Pyrethroids on Calcium and Oligophosphate Turnover in Nerve Cells

The search for corroborating evidence from biochemistry was based on the findings that micromolar amounts of the sodium channel effector DDT have strong inhibiting properties on membrane bound NaK-Mg-ATPase and oligomycin- sensitive Mg-ATPase [747, 748], which is the terminal enzyme of oxidative phosphorylation [749]. Some Ca-ATPases in the lobster axon are even inhibited by nanomolar amounts of DDT [750]. Since nerve-ATPase also play an important role for the proper course of action potential as well as for the functioning of Na/K-pumping, the connection ATPase and Na-channel deserves a thorough analysis. It could be shown indeed, that some type-I pyrethroids are also inhibitors of nerve-ATPases [751, 752, 753, 756]. Pyrethrin I and its very close variation allethrin block only the not-temperature-sensitive Ca-ATPase, which needs Na and K for optimum stimulation by Ca and fixes Ca on the cytosolic surface of the membrane. It acts as a Na-Ca-protein kinase-phosphatase [754] and regulates the permeability of the membrane for Ca into the inner part of the nerve cell, while the α-CN-pyrethroids have a profound effect on Ca-Mg-K-ATPase [757, 758], which effectively keeps the plasmatic concentration of free Ca low by sequestering Ca or by pumping the Ca out of the cell or into Ca collecting mitochondria within the cell. This ATPase thereby keeps a concentration gradient of Ca of 5 orders of magnitude, 10^{-8} M in cytoplasma, 10^{-3} M in mitochondria or outside the cell. Here again, permethrin with 10^{-8} M plays an intermediate role in being less selective, acting on both ATPases. More recent results [759] suggest that the action of pyrethroid on Ca-ATPase is unspecific, leading to the inhibition or stimulation, depending of the chain length of the anular lipids of this membran bound protein. Moreover, these effects are relatively non-stereospecific.

No effect of allethrin and permethrin was seen on ouabain-sensitive Na-K-ATPase, nonmitochondrial Mg-ATPase and axoplasmatic ATPase [758]. Ouabain is another agent that causes elevation of internal Na like pyrethroids, but by inhibiting the ATPase responsible for pumping out the Na.

Thus the inhibitory effect on Ca-uptake [755] or inward flux by blocking Ca-ATPase leads ultimately to nerve blocking, while inhibition of the removal of Ca from cytoplasm results in an excitatory effect on nerves by the stimulation of transmitter release [756].

The concentrations needed to have effects on this ATPase, however, are much higher than so far observed for certain nerve preparations (pI_{50} 15–18). On synaptosomes cypermethrin showed a pI_{50} of 8, permethrin a pI_{50} of 7.

Some hints to the actual binding regions [760] come from experiments using carefully and partially delipidized ATPase preparation from lobster [761].

Although these preparations lose their ATPase activity to a large extent, they are, in respect to relative inhibition of ATP cleavage, much more permethrin-sensitive than normal ATPase. The important of Ca also for the Na-channels is shown by the fact, that agents, affecting Ca-regulation are also effective in Na-channel performance, but not in the resistant insect [702].

Only the insecticidal isomer of fluvalinate is a Ca-blocker like verapamil [762]:

fluvalinate

verapamil

Both allethrin and fluvalinate are very effective inhibitors of acetylcholin receptor activated Ca uptake [763].

The inhibition of Ca-uptake alone apparently has no direct relation to insect toxicity [764]. It depends on an optimum lgP value of 6.6. In regard to the stimulation of neurotransmitter release the effect of resmethrin in rat-brain on Ca-metabolism is more important than on the Na-channel target site, in contrast to cypermethrin [760].

In addition, phosphordiesterase was also directly affected by permethrin. However, the concentrations observed so far are again much higher than needed to affect Na-channels. Other effects on important steps in Ca turnover were also demonstrated by the inhibition of calmodulin by DDT, permethrin and, to a lesser extent by cypermethrin [764]. Since calmodulin plays a vital role in many regulatory processes, e.g. phosphorylation of membrane proteins by Ca-dependent protein kinases, this aspect was also directly investigated. The activation and inactivation of several ion channels by phosphorylation is one common general mechanism. Studies with 10^{-13} M deltamethrin in Retzius cells of the leech (a nerve preparation, which readily shows repetitive activity with *all* pyrethroids) showed, that the very regular spontaneous depolarizations, the repetitive firing of the nerve, is caused by the involvement of an oscillating ion current in and out of the cell, controlled by intracellular Ca and mediated by protein kinases [766]. 10^{-13} M deltamethrin stimulated the synaptosomal activity of cAMP dependent, Calmodulin activated protein kinase up to 200% [825].

A regularly occurring phosphorylation and dephosphorylation is one explanation:

calmodulin,
c GMP
or diacylglycerol

kinase

ATP + [channel protein]   [channel protein]—O—P—O$^{\ominus}$

phosphatase

The same effect, as with pyrethroids on this and other nerve preparations, is accomplished with phorbolesters [767], an activator of Ca-dependent protein kinase C which controls inter alia the release of neurotransmitters by regulation of Cl, K and Ca channels. This protein kinase switches by a slight increase of intracellular Ca from cytoplasma to the inner membrane and is normally turned on by phospholipids and diacylglycerol, constituents or fragments of the nerve membrane.

Nerve membranes of typ I pyrethroid resistant strains of certain insects show reduced sensitivity to Ca [768a]. A substantial increase in the GMP-level, but not in cAMP, in *Mamestra* moth was observed after treatment with permethrin [769].

The oscillating phosphorylation and dephosphorylation mentioned above certainly dissipates energy. Whether the increase of normal oxygen consumption (45 μM/mg/h, consuming 90 μval electrons in resting insects [770]) in permethrin poisoned insects [771] or the increase in body temperature in cismethrin poisoned mice [772] is connected with this waste of ATP remains to be seen.

7.3 Involvement of Pyrethroids in Oxidative Processes

A step further in the direct involvement of pyrethroids in oxidation processes is the surprising observation that the insecticidally active isomer of the DDT-pyrethroid-hybrid GH 380 is an inhibitor (K_i 7.10^{-8} M) for a specific lipophilic mixed functional neural oxidase, while the inactive one is not [125] (Table 103). However, the link to insect toxicity is apparently not a causative one.

Table 103. Inhibition of induced blowfly microsomal MFO by active and inactive isomers of pyrethroids [125]

		Insecticidal activity	Concentration mM	MFO Inhibition konst. K_i mM
X	isomer			
H	S	Active	0.05	0.07
	R	Inactive	2.0	No inhibition
C≡CH	S αR	Active	0.01	0.002
	R αR	Much less active	0.01	No inhibition
	S αS	almost inactive	0.05	0.014
	R αS	Inactive	0.01	0.003

Also a phenol oxidase from *Spodoptera* was inhibited by less than micromolar amounts of deltamethrin [773]. A patent claims pyrethroids as insecticides and antagonists of oxidative metabolism [774].

The reason for the synergism of deltamethrin and chlodimeform is the inhibition of monoaminooxidase by the pyrethroid, thus leading to an increase of octopamine and tyramine [832].

A concentration of 0.5 ppm permethrin decreased the activities of cytochromoxidase and lactatdehydrogenase in the snail *Lymnaea acuminata* [833]. The inhibition of the microsomal Cytochrom P450 activity by deltamethrin in rats is presumably associated with the disruption of synthesis of microsomal proteins, because the pyrethroid had no direct effect on the microsomal enzymes in vitro [834].

7.4 Effect of Pyrethroids on Postsynaptic Cholinergic Na-channels

The overall effect of pyrethroids on a single nerve is the facilitation of axonal postsynaptic conduction of the nerve signal by greatly exaggerating the necessary sodium influx into the cell through pyrethroid modified sodium channels. This leads to excessive neurotransmitter release, e.g. acetylcholine, through the synaptic cleft to neighbouring cells. However, certain pyrethroids at more elevated concentrations also have specific effects on the postsynaptic nicotinic acetylcholine receptor complex, part of which is a sodium channel, opened by the binding of acetylcholine. Allethrin, resmethrin and tetramethrin inhibit rapidly and non competitively ($pI_{50}6$), the binding of histrionicotoxin (HTX), a typical lipophilic alkaloid from frogs, which modifies that channel in the activated, open state by binding at a different site with a marked negative temperature coefficient [763, 775, 777, 778]. This binding is competitive to imipramine. On the other hand, the HTX binding is stimulated by these pyrethroids in the non activated closed-channel receptor state.

Both enantiomers of cis fenfluthrin are active on that nicotinic Na channel at 10^{-6} M [842].

The more toxic α-cyano-pyrethroids act much more slowly and less specifically. Therefore, there seems to be no important contribution to the overall toxicity symptoms in insect or rat from this target site.

The overall effect of pyrethroids on the receptor-gated Na-channel, however, would be the stimulation of a electric potential flux through the ganglion, or the facilitation of nerve conduction.

Some transient symptoms of the intoxication of rats with pyrethroids are of typical cholinergic nature. Since experimental rats have to take up a larger dose of less toxic pyrethroid as compared to poisoned insects, fast appearing effects can be brought about by this higher internal concentration as a pharmacokinetically controlled transient effect.

In rat brain cortex in vivo there is even a 50% inhibition of acetylcholine esterase by 250 mg/kg permethrin within 30 min., which can be reversed by acetylthiocholine [779].

On the other side, the insecticidally active isomers of deltamethrin, cypermethrin, and resmethrin induce the activity of bovine-acetylcholine esterase, while the inactive are inhibitory [667].

7.5 The Effect of Pyrethroids on Gabaergic Chloride Channels

In the toxicological symptoms of poisoned rat, there is a clear component typical for disturbances in the gabaergic system (hyperactivity, tremor, tonic seizures). The related basical effect can also be found on the molecular-toxicological level. The doses are higher than the ones to affect Na-channels, though. But the rat is treated with much higher doses as compared to insects, so again, high concentrations may be sufficient for producing this visible syndrome.

Deltamethrin (10^{-7} M) and permethrin stimulate gaba release from guinea-pig synaptosomes [600]. It was found, that only the insecticidally active α-cyano-pyrethroids, type II pyrethroids, have effects on the gaba binding postsynaptic chloro ionophore, the chlorid channel, a transmembrane protein with binding sites for regulating benzodiazepins like diazepam, and toxins like t-butyl-bicyclophosphate (TBPS), picrotoxinin, avermectins and kainic acid [780, 781, 522]. The Chloride channel counterregulates the effects of the sodium channel. While the latter facilitates nerve conduction by its activation, the former inhibits the overall nerve activity. Deltamethrin (10^{-7} M), > cis > trans-permethrin, decreases the binding of kainic acid in mouse brain. In the same order these compounds produce the above mentioned neurotoxic symptoms [782].

The action of pyrethroids on isolated muscle fibers from the neuromuscular junction in crayfish is similar, but slower than that of the gaba-antagonist picrotoxinin. This effect can be diminished by benzodiazepins, which enhance gaba binding and the resulting chloride current into the nerve cell [783], leading to an inhibition of nerve signal input. In addition only the insecticidally active α-cyano-pyrethroid inhibits non-competitively the binding of another gaba antagonist TBPS [781, 784, 785] in brain synaptosomes, but not the binding of picrotoxinin [786].

The inhibition of TBPS-binding in mice of a set of α-cyano-pyrethroids is correlated with toxicity to mice [775]. Deltamethrin has no voltage depending effect on Cl-channel [787].

In rat brain microsacs the gaba-induced Cl-flux into the cell is inhibited by 1-R-cis, α S > 1-R-trans-α S-cypermethrin, fluvalinate, allethrin [788], lindan and polychlorocyclodien in micromolar concentration similar to TBPS.

Conflicting evidence has been found in experiments with rat brain synaptosomes, where cycloprothrin enhances dose dependent gaba binding, TBPS- and diazepam binding as well [66], although in quite a high concentration of 100 μM.

The analysis of input-resistance in crayfish muscle fiber after pyrethroid intoxication seems to confirm the involvement of the gaba receptor complex in the type II syndrome [789, 790]. However, in nerve preparations with TTX blocked Na channels no effect was seen with 10^{-10} M deltamethrin on gaba binding. Thus, the Na channel is still the most pyrethroid sensitive part in the nerve system. In the crayfish stretch receptor neuron the Na channels are a million times more sensitive to deltamethrin than the gaba receptor complex [789a]. It remains to be shown, that the gabaergic component in pyrethroid toxicity as mentioned above has also some importance in insects. These invertebrates certainly possess gaba-receptors [787, 790b], however with a different pharmacology than the ones in vertebrates.

The synergism [790a] of pyrethroids and avermectin, a strong stimulator of gaba binding and release as well [790c], being about as insecticidally active as deltamethrin, does not support the relevance of the mode of action for pyrethroides at the Cl-channel of the insects or vice versa, the modes of action in fashion of avermectin.

In summary, an inhibiting effect on gaba binding by pyrethroids would result in a reduction of the inward Cl-current, which antagonizes the inward Na-Current, thus enhancing the overall nerve conduction in the ganglion.

The extremely high nerve activity and insecticidal activity of pyrethroids may nevertheless result from a mutual potentiation of Na- and Cl-channel effects [66], by the disturbance of the buffer role of inhibitory gabaergic neurons and excitatory cholinergic, dopaminergic, glutaminergic and other neurons.

7.6 The Effect of Pyrethroids on the Release of Transmitter and Hormons

The insect nervous system is tightly protected by the perineurium, functioning like an effective blood-brain-barrier. But there are parts of the nervous system, the neurosecretory cells, which are in direct contact with the hemolymph. A poison in the hemolymph is likely to arrive at this neurohaemal area of the nervous system first (however, see p. 87). In fact, these cells are extremly sensitive to permethrin [791]. Seven min. after application of 10^{-11} M permethrin to such a nerve preparation a violent, very long lasting increase of the normal repetitive activity starts due to action on impulse generating sites in the membrane. Deltamethrin developed these effects more slowly. The effective concentration in doubling the normal frequence after 40 min. was as follows:

Deltamethrin 10^{-10} M
Bioresmethrin $2 \cdot 10^{-10}$ M
Permethrin 10^{-9} M
Bioallethrin $2 \cdot 10^{-7}$ M

After application of an LD_{95} on *Rhodnius* (0.057 µg/g) up to a 300% increase of nerve activity resulted within 15 min. This hyperactivity may well be the start for a dramatic Ca-depending exocytosis of neurohormones into the hemolymph, for secondary effects like increase of trehalose in hemolymph [792] after direct action on corpus cardiacum, or for heavy loss of water, leading to a strong reduction in volume of hemolymph after release of diuretic hormones [793]. This loss of water may be the actual cause of death [843]. By the same token, hyperlipemic and adipokinetic [794] hormones are released. The very first effect, before hyperexcitation of the locust appears, is a 160% increase of the typical lipoprotein in the hemolymph as fuel for escape by flight [795].

Conversely it was found in rat-neurohypophysis, that deltamethrin and resmethrin inhibited the release of hormones after stimulation [796]. Target sites in the neurohemal tissue are Na-channels, since ultrastructural alterations of mitochondria in isolated corpora cardiaca cells, caused by micromolar amounts of bioresmethrin, could be prevented by petreatment with µm TTX [797].

Related to the exocytosis of hormons is the release of neurotransmitters into the synaptic cleft, observed for example with micromolar concentrations of deltamethrin, causing the specific increase of noradrenaline release in stimulated sympathetic nerves in the rabbit heart [798]. Only α-cyano pyrethroids enhance Ca-dependent, K-stimulated release of norepinephrine from rat brain synaptosomes (10^{-9} M deltamethrin) [798a].

Similarly, micromolar concentrations of fenvalerate induce after a short time in nigrostriatal preparations from the rabbit the release of dopamine and acetylcholine, which could be reduced by micromolar Ca-antagonist nitrendipine. This release was abolished by micromolar TTX [790].

All this can be explained by the opening of Ca-channels as a secondary effect of membrane depolarizations caused by the action on presynaptic Na-channels.

However, a concentration of 10^{-8} M pyrethroids stimulates the release of inhibitory transmitter gaba from synaptosomes [800] independent of external Ca. The concentration-dependent effect was produced by deltamethrin > > > permethrin > DDT > veratridine.

Presynaptic nerve terminals of deltamethrin-treated houseflies showed a marked depletion of synaptic transmitter vesicles [755]. The substance, which is toxic to other flies, released by deltamethrin-treated, prostrate flies, was shown to be tyramine and its metabolite octopamine. Both are activators of adenyl cyclase. As a result elevated levels of the second messenger cAMP are produced [755a].

In one leaf eating ladybird, cypermethrin in low doses stimulated the development of ova, induced oocyte maturation and oviposition due to the release of two endocrinous factors, one of them being juvenile hormone [800a]. This could be a part of the mechanism of the resurgent effect in certain insect populations after a pyrethroid treatment.

7.7 Various Effects of Pyrethroids on Other Nerve Membrane Targets

In addition to the secondary effect of neurotransmitter- and hormone release, stimulated by the action of pyrethroids on the sodium channel, another secondary effect of pyrethroids is the simulation of a phospholipase [801].

However, this effect is caused by a number of diverse effectors which cause increase of internal sodium, either by opening Na-channels, by inhibiting NaK-ATPase, or by providing additional Na-passages through the membrane (ionophores). The inward Na-current triggers a cascade of secondary events, starting with the activation of phospholipase which cleaves the membrane constituent phosphatidyl ionositols, thereby producing two second messengers: inositol phosphate, which releases Ca from internal stores, and diacylglycerols, which activate protein kinase C. If this cascade goes on and on, toxicological relevant effects are certain.

All the previously mentioned established, putative and proposed fundamental molecular-pharmacological targets for pyrethroids have their location in nerve cell membranes in common. These lipid membranes are highly anisotropic by virtue of the physical nature of of their phospholipids (fatty esters of a glycerol phosphate) as being natural detergents, and in addition, due to the high chirality conferred by the R-configuration at carbon 2 of the asymmetrically substituted glycerol, especially in the polar headgroup. Great disturbance of packing order, transition temperature for different liquid-crystalline phases and viscosity are caused by minute amounts of certain compounds, e.g. lysolecitins. Pyrethroids like cis and trans permethrin, cypermethrin, fluvalinate, and fenpropathrin (allethrin much less so), also cause a decrease of packing order anisotropy of lymphocyte membranes in micromolar concentrations [802].

They also alter gel-fluid phase-transition profiles only [803] and lower the phase-transition temperature. There is no change in the actual fluidity [804].

Lowering the packing order of the lipid membrane changes the radius of curvature of the lipid bilayer, important for lipid-protein interaction. Since the lipophilic pyrethroids preferentially should stay in the most hydrophobic part of the membrane, they disturb the packing order in the lipid bilayer core [805].

However, analysis of the fluorescence spectra of the permethric-ester of pyrenemethanol (a pyrethroid with moderate insecticidal activity [72]) in lipid bilayers allows the location of the ester moiety in the lipid-water interface of the glycerol backbone. The pyrethroid molecule adopts thereby the folded conformation as mentioned before (page 78), and uses four lipid molecules for binding [804].

The effect of permethrin on these membranes is enhanced by calcium [806].

The importance of the effect of pyrethroids on the lipids in the membrane is shown once more by experiments with artificial liposomes of several phospholipids having taken up fully functioning nerve membrane fragments such as Na-channels. These liposomes interact with the 1 R-tetramethrin-isomer in a manner expected for proper sodium channel systems, which can be blocked by TTX. The 1-S-isomer however, also causes a sodium inward current of the same magnitude, which could not be blocked by TTX. This effect is not observed in the natural neural phospholipid system [807].

So far however, the concentrations needed to cause these bio-physical effects seem to be too high to be of toxicological relevance for insect poisoning.

D. Pyrethroids as Commercial Insecticides

8 Pyrethroids as Commercial Insecticides

Right after the synthetic pyrethroids of the Rothamsted-Laboratories of M. Elliott and coworkers, and those from Sumitomo were found to be most interesting for agricultural use on a large scale due to their very high activity, photostability, low acute mammalian toxicity (Table 104), as well as superior and cheaper seasonal performance in comparison to existing standard insecticides (Table 105) great efforts were made by many companies to get hold of patent rights. Within a short time

Table 104. Comparative toxicities of pyrethroids and older standard insecticides [809]

Insecticides	Abs. Toxicities [816, 665]		Rel. Toxicities		Therapeut. Index
	Am. cockroach		housefly	cotton pests	$\dfrac{LD_{50} \text{ Rat}}{LD_{50} \text{ Fly}}$
	LD_{50} topical ppm	LD_{50} injected µg/g			
Organochlorines					
DDT	10–30	5–18	1	1	11
Lindane	4–7	3–7			
Dieldrin	1.3	1.1			
Organophosphates					
Malathion	24	8.5	0.2		50
Parathion	1.2	1	3	1.5	9
Diazinon	2.0	0.8	2		
Dimethoate			5		390
Carbamates					
Carbaryl				1	
Rothenone	2000	5–8			
Pyrethroids					
Pyrethrines			1		74
Bioresmethrin	1.5	1.1			
Permethrin			5	10	
Cypermethrin			10	25	
Deltamethrin	0.04	0.05	=100	=100	5300
Fenvalerate			4	20	

Table 105. Application of pyrethroids in cotton against *Heliothis virescens* [810]

Pyrethroid	Application rate	Number of applications	Seasonal application rate	Cotton seed yield
	(g/ha)		kg/ha/season	(kg/ha)
Fenvalerate	150	11	1.6	2500
Cypermethrin	100	10	1.0	2530
old standards:				
Toxaphen	1000 ⎫		21	
DDT	500 ⎬	21	10	1220
Parathion	750 ⎭		15	

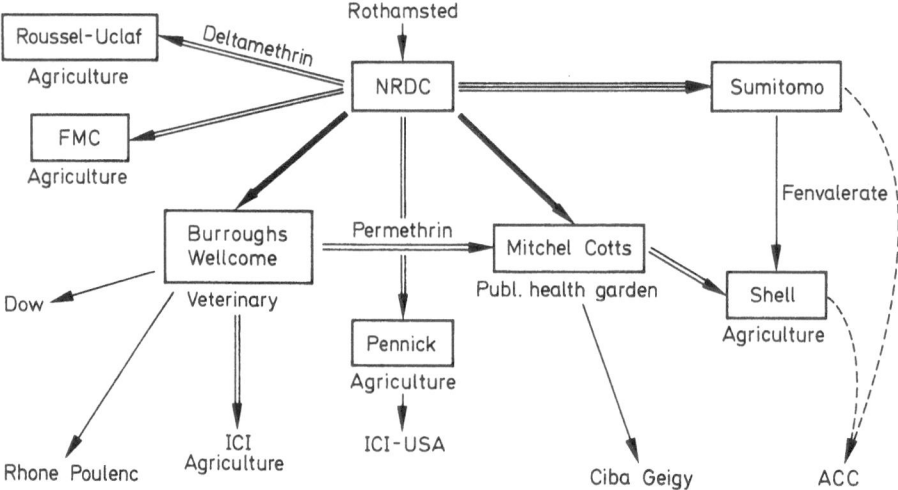

Fig. 33. Network of licensing and sales rights for the photostabil pyrethroids invented at Rothamsted and Sumitomo's as negotiated by the National Research and Development Corporation (NRDC) and Sumitomo Chemical Corp. with a group of seven companies

a complicated network of licensing and interdependencies was knit together, which comprised active ingredients, preproducts, intermediates and processes. The network of licence and sales rights for the NRDC-pyrethroids resmethrin, cypermethrin and deltamethrin and Sumitomos fenvalerate (status 1984 according to Wood-Mackenzie reports [808] are as shown in Figure 33.

NRDC-rights: Shell: whole world except Japan and USA
              ICI: whole world except Japan
              Sumitomo: Japan
              FMC: whole America

Roussel Uclaf: Germany, France, Italy, Benelux, Eastern Europe, Asia (except Japan), Francophon Africa

Mitchell Cotts: whole world except Japan, USA, Canada

In the meantime the general interest in this research intensified up to 1982. This effort does not show up in scientific journals, but shown by the smaller or larger number of patent applications filed by the chemists and entomologists of the following research institutions (see Vol. II of this book): American Cyanamid (USA); Asahi Chemicals (Japan); Bayer (Germany); BASF (Germany); Chinoin (Hungary); Ciba-Geigy (Switzerland); CSIRO (Australia); Danippon (Japan); Dow (USA); Dynamit Nobel (Germany); Ethyl Corporation (USA); FMC (USA); ICI (UK); Indian Explosive (India); Katsuda (Japan); Kuraray (Japan); Mitsubishi (Japan); Mitsui Toatsu (Japan); Mobil Oil (USA); Montedison (Italy); Nippon Kayaku (Japan); Nippon Soda (Japan); Nissan (Japan); Rothamsted Institute (UK); Roussel Uclaf (France); Sagami (Japan); Sankyo (Japan); Schering (Germany); Shell (USA, UK Netherlands); Stauffer (USA); Sumitomo (Japan); Yoshitomi (Japan) and Zoecon (USA).

The new pyrethroides were readily accepted by the market. In the beginning the demand exceeded the supply capacity, which grew rapidly by tremendous scientific and technical effort (Fig. 34). The innovations in this field apparently were accomplished at the expens of other fields of agrochemical research (Fig. 35).

The production capacity in 1982 was 4000–6000 annual tons of permethrin equivalents, worth 2 billion US$. By 1985 the 12000 annual tons of permethrin equivalents corresponded to 15–20% of the foliar insecticide world market [811]. Other estimates give marketshares for the pyrethroides in 1985 of 23.7%, for 1986 25.1% [813] and forcast for 1990 about 35%. However, the very optimistic expectations in the beginning could not be fullfilled, as shown by the up and down and inconsistencies of the forcasts in the Table 106. In 1987 Cypermethrin was on rank 2 in the list of insecticides for the world market, fenvalerate on rank 6.

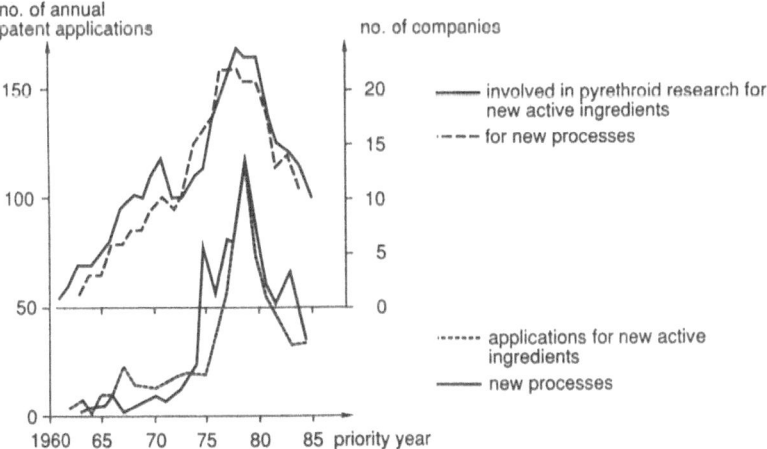

Fig. 34. Intensity of pyrethroid research for new active ingredients and processes as determined by number of patents and institutions involved

Table 106. Forecast and market developement for pyrethroids on the insecticide world market [808][a])

Forecast				Market share achieved			
Year	%	Value $ million	Annual growth rate	Year	%	Value $ million	Annual growth rate
1985	30			1976—80			55%
1986	16	1300	15—18%	1978 (USA)	19		
1986	21	1750		1980	9	350	
1986	30			1981	[b])		
1986	15	1100		1982	12	500[c])	
1986			12.7%	1983	16		
—90				1984	20	1140 (2000)[d])	
1990	30			1985	[e])	1200 (1500)	
	(35)	2150	11.8%	1985 (USA)	50% in cotton	120	
1990	20	1400	7%	(USSR)	30% in cotton	70 70	
				1986	[f])	1375	
				1987	47% in cotton	1140	
1995	56			1988		1150[g])	
				1988	4000–5000 to per- methrin equiva- lents		

[a]) Global pyrethroid market in 1987 [837]: cotton 46.5%, fruit and vegetables 21.6%, other crops 16.7%, animal 9.3%, other non agricultural outlets 5.9%;
[b]) 230 insecticidal active ingredients are listed in the Farm Chemical Handbook, 12 are pyrethroids (1981) [812];
[c]) Corresponds to 33 million hectares treated with pyrethroids. Forecast for 1986: 50 million ha;
[d]) This is the 3rd largest group on the agrochemical world market for insecticides;
[e]) In 1984 190 insecticidal active ingredients are listed, 20% are pyrethroides [813] including isomer enriched ones. About 80 different organophosphates are on the market;
[f]) In 1986 14 different pyrethroids (not including isomer enriched modifications of the same structure) are on the agro market
[g]) NRDC pyrethroids only: 820 Mio $

Nevertheless, the pyrethroids will certainly find their definite larger, but limited share of the insecticide market of finally between 30 and 35% around 1995.

However, in some respect this group of new compounds reached in the meantime already the strange status of "patented propriatory commodity". This is due to the steadily decreasing prices — by 25% between 1982 and 1986 [814] — apparently originating from the complex system of ownerships, from technical improvements, increased activity and increased number of highly active newcomer active ingredients; but also due to stagnating markets. Thus, the price for "synthetic pyrethroids" were as follows: 54.40 $/lb in 1985, 51.20 $/lb in 1986, 48.80 $/lb in 1987 [820].

The pyrethroids are here to stay, provided they are reasonably used in connection with other insecticides and pest management practices in order to avoid widespread pyrethroid resistance. This resistance may come about in an alarming speed and may threat large parts of the cotton growing area, like witnessed in the Esmerald Valley in Australia in 1983/1984. Not to forget the lessons to be heeded from the history of resistance to DDT, and connected with this, the fate of former cotton growing areas. In the meantime the international insecticide industry is fully aware of the potential earnest of the situation. The Insecticide Resistance Action Committee (IRAC) is outlining a set of actions, comprizing close monitoring of resistance levels with standardized methods and proposals for counter measurements in case of insect pest controll failures.

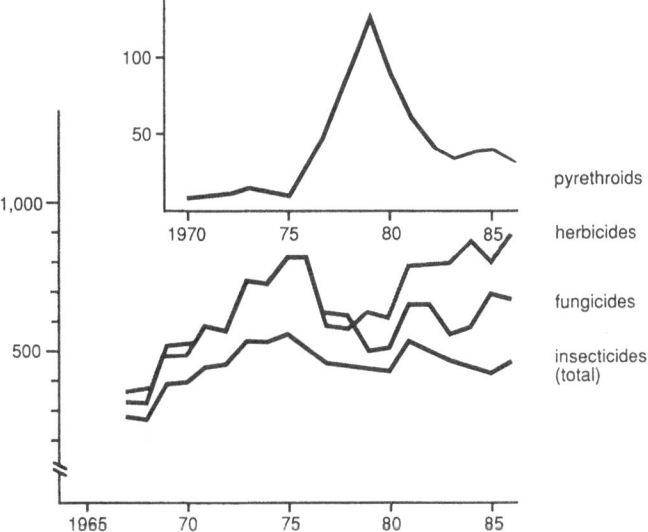

Fig. 35. Pyrethroid research and research for new active ingredients for plant protection (number of annual patent; year of priority)

In 1987 the following companies have been owners of propriatory commercial pyrethroids other than the affore mentioned ones: American Cyanamid, Bayer, FMC, ICI, Mitsui Toatsu, Roussel Uclaf, Zoecon/Sandoz, Sumitomo, Nippon Kayaku.

Commercial synthetic pyrethroids for general agricultural application against insects are the following compounds: alphamethrin, betacyfluthrin, bifenthrin, cycloprothrin, cyfluthrin, cyhalothrin, cypermethrin, deltamethrin, esfenvalerate, etofenprox, fenpropathrin, fenvalerate, flucythrinate, fluvalinate, lambdacyhalothrin, permethrin, tefluthrin, tralomethrin. For formulae and other names see Table 112 p. 192ff.

For a general orientation the approximated median conversion factors of relative application rates of the most important multipurpose pyrethroide are shown in the following correlation Table 107.

Table 107. Median conversion factors of relative application rates of most important pyrethroids

	Deltamethrin	Cis-Cyhalothrin	Alphamethrin	Esfenvalerate	Cyfluthrin	Cypermethrin	Flucythrinate	Bifenthrin	Fenvalerate	Permethrin	
Deltamethrin	1										
Cis-Cyhalothrin	1	1									
Alphamethrin	1.5	1.5	1								
Esfenvalerate	2	2	1.3	1							Increasin
Cyfluthrin	2	2	1.3	1	1						activity
Cypermethrin	4	4	2.7	2	2	1					= lower
Flucythrinate	4	4	2.7	2	2	1	1				application
Bifenthrin	4	4	2.7	2	2	1	1	1			rate
Fenvalerate	7	7	4.7	3.5	3.5	1.8	1.8	1.8	1		
Permethrin	10	10	6.7	5	5	2.5	2.5	2.5	1.4	1	

Commercial pyrethroids and application rates usefull for special pest problems:
a) spider mites:
   Biphenthrin (40–80 g/ha), lambdacyhalothrin, fenpropathrin, (50–100 g/ha), flucy-thrinate (100–150 g/ha), fluvalinate (100–200 g/ha), (NCI 85193 (50–100 ppm, experimental acaricide)
b) Pyrethroids having low fish-toxicity, therefore potentially usefull for application against rice pests:
   cycloprothrin, ethofenprox (MTI 800; experimental) HOE 498 (experimental)
c) Commercial pyrethroids for special purpose:
   fenpropathrin (white fly, spider mites)
   fluvalinate (*Heliothis*, spider mites)
   tefluthrin (*Diabrotica* in soil)
d) Commercial pyrethroids for veterinary use (decreasing order of activity) against cattle tick:
   flumethrin, cyhalothrin, cypermethrin, cypothrin, deltamethrin, fenvalerate
   symbovine flies:
   cyfluthrin, cyhalothrin, cypermethrin, fenvalerate, flucythrinate, permethrin
e) Commercial household pyrethroids:
   pyrethrin, allethrin, benfluthrin, bioallethrin, bioethanomethrin, cyphenothrin, cypermethrin, deltamethrin, furamethrin, kadethrin, kikuthrin, permethrin, pheno-thrin, resmethrin, bioresmethrin, tetramethrin, vaporthrin, prallethrin, ethofen-prox.

The following tables show some application rates of commercial pyrethroids under practical conditions.

Table 108. Application rates in cotton 1979 [815]

	Spain		Mexico
	absolute	relative	relative
Deltamethrin	12.5–18.8 g/ha	1–1.5	1
Cypermethrin	100–150 g/ha	8–12	4
Fenpropathrin	100–150 g/ha	8–12	
Permethrin	125 g/ha	10	8
Fenvalerate	150–180 g/ha	12–15	6

Table 109. Application rates of pyrethroids in different crops (g/ha; %) 1985

Common name	Cotton		Vegetables	Fruits	General agriculture
	Normal	in Egypt			
Deltamethrin	10– 20 g	45 g	5– 25 g	0.00075–0.0018%	5– 18 g
Cypermethrin	50–150 g	144 g	50–100 g	0.001–0.006 %	40–150 g
Permethrin	30–200 g		30–100 g	0.0025–0.015 %	150–200 g
Fenvalerate	20–100 g	288 g	25–150 g	0.0025–0.015 %	100–150 g
Cyfluthrin	25– 35 g	90 g	12– 25 g	0.0025 %	25– 35 g
Cyhalothrin	25– 36 g	72 g	0.0025–0.005%	0.0025 %	50– 70 g
Cis-Cyhalothrin	12– 18 g	45 g			5– 30 g
Alphamethrin	7– 35 g	60 g	5– 30 g	0.001–0.006 %	7– 35 g
Esfenvalerate	25– 35 g	72 g		0.001–0.003 %	5– 25 g
Tralomethrin	15– 20 g		15– 20 g	15– 20 g	15– 20 g
Fenpropathrin	50–100 g	360 g		0.01–0.02 %	100–150 g
Bifenthrin	20–100 g		25– 75 g	0.0005–0.006 %	25–100 g
Flucythrinate	40– 75 g		30– 75 g	0.0025–0.0075 %	40– 70 g
Fluvalinate	25–170 g		56–170 g	0.0025–0.02 %	50–150 g
Ethofenprox	100–200 g		75–150 g	0.0075–0.02 %	75–200 g

Table 110. Recommended use rates for Cypermethrin [817, 818]

Main crop damaging insect	Crop	Rate (g/ha)
Highly susceptible species:		
Diabrotica, Sitona, Leptinotarsa	Vegetables	25–50
Macrosiphon, Metapolophium	Corn, cereals	15–25
Amblitropodea	Pasture	15–25
Susceptible species:		
Heliothis (a., z., v.)		50 (Spain, Mexico)
Plusia, Spodoptera, Trichoplusia }	Cotton	–100 (Australia,
Bucculatrix, Pectinophora		South Africa)
Myzus, Aphis		
Plutella, Pieris, Mamestra }	Vegetables	50–100
Empoasca, Trialeurodes		

Table 110. (continued)

Main crop damaging insect	Crop	Rate (g/ha)
Less susceptible species:		
Empoasca, Aphis, Bemisia, Dysdercus	Cotton	100–150
Busseola	Corn	100–125
Ceuthorrhynchus, Dasyneura	Rape	200–250

Recommended dosage rates for alphamethrin [551a]

Order	Pest	Dosage rate (g a.m. per ha)
Coleoptera	Weevils, flea beetles, beetles	7.5–30
Diptera	Leaf miners, midges, maggots,	10–30
	Mediterranean fruit fly	17.5–30
Hemiptera	Aphids, scales, whiteflies, leafhoppers, plant hoppers	10–30
Homoptera	mealy bugs, suckers, leaf rollers	
Hemiptera	Plant bugs, capsids, stainers	7.5–30
Heteroptera		
Hymenoptera	Sawflies	17.5–30
Lepidoptera	Stalk borers, bollworms, cutworms,	5–35
	armyworms, loopers, leaf rollers	
Orthoptera	Grasshoppers	7.5–20
Thysanoptera	Thrips	10–30

Table 111. Recommended use rates of fenvalerate [819]

Main crop damaging insect	Crop	Rate (g/ha)
Very susceptible species:		
Spodoptera, Pieris, Noctuidae,	Vegetables	50–100
Plutella, Trichoplusia, Myzus pers.		
Schizaphis, Aphis sacchari,	Sorghum	50–100
Epilachna, Hypera	Alfalfa	
Nuridae, Nabidae, Cicadellidae, aphids		
Heliothis, Plathypena, Grapholita	Soybean	
Susceptible species:		
Leptinotarsa, Chrysomelidae, aphids,	Vegetables	
Busseola, Spodoptera,	Corn	100–200
Marasmia, Heliothis, Pectinophora,	Potatoes	
Spodoptera, Anthonomus, thrips	Cotton	
Less susceptible:		
Dysdercus, Bucculatrix	Cotton	200–400
fruits moths, budworms, aphids	Fruits	0.01–0.015%

Developmental and commercial active ingredients are characterized for the public use by code numbers, common names and, as formulated preparation, by the registered trade name. Together with the CAS registry number these names are the main keys for a computer recherche for further information.

Since pyrethroids as aliphatic species usually can show up in the form of stereoisomers, it is of importance to characterize the pyrethroid in question by its stereoisomeric composition. Only some of the isomers are biologically active. For comparison of the real insecticidal potency of different trade products their content of active isomers has to be taken into account. The following final table lists all the contemporary (1989) trade pyrethroids and most promising developmental products with the above mentioned characteristics together with their institution of invention and industrial production. The number of commercial pyrethroids seems to level off, and in connection with this, the number of additional common names. However, the number of trade names of pyrethroid preparation certainly will keep growing in the next decade together with a still growing market. In some cases, the trade names in the table are not put in connection with the basic producer. In these cases it merely lists the trade names in connection with a common name.

8.1 Compilation and Characterization of the Developmental and Commercial Pyrethroids

Table 112. Commercial, developmental and experimental pyrethroids

Entry No.	Formula	Stereo-chemistry	No. of isomers	Content of active isomers %	CAS registry No.
1		1R trans; S	1	100	121-21-1
1a		1R trans, E; S	1	100	121-29-9
2		1R trans; S	1	100	25402-06-6
3		(±) cis/trans, RS	8	25	91103-69-4
4		(±) cis/trans; RS	8	25	584-79-2
5		1R trans; RS	2	50	
		1R trans; S	1	70–100 100	28434-00-6
6		(±)	2	50	15589-31-8

Common name	Code No.	Inventing institution	Basic producer* distributor	Trade names
Pyrethrin I		Nature	Nature* (see p. 3)	Pyrethrum Pelitre
Pyrethrin II		Nature	Fairfield American McLaughlin Gormley-King	Keatings powder Pyrenone Pyrocide, Spruzit
	ENT 7543	Nature	FMC Pennick Synarome Corp.	Pyrethrolone
Cinerin I		Nature	Prentiss Drug	Prentox
		Chinoin		
Allethrin		Boyce-Thomson	McLaughlin-Gormley-King* Sumitomo* Fairfield American*	Pynamin Alleviat
Pallethrin			Procida* (Roussel-Uclaf)	
d-trans Allethrin			McLaughlin-Gormley-King*	D-Trans
d-Allethrin			Sumitomo*	Pynamin forte
Bioallethrin	RU 22366		Procida*	Bioallethrin
Esdepallethrin	RU 3054	Roussel Uclaf	Roussel Uclaf*	Raidsect Esbiothrin Esbiol
Terallethrin	M 108	Sumitomo		

Table 112. (continued)

Entry No.	Formula	Stereo-chemistry	No. of isomers	Content of active isomers %	CAS registry No.
7		(±)	2	50	51388-35-3 98568-88-8
		(+)	1	100	98568-89-9
8		(±) cis/trans 1R cis/trans; RS	8 4	25 50	23031-39-9
8a		(±) cis/trans	4	50	70-38-2
9		(±) cis/trans	4	50	70-43-9
10		(±) cis/trans 15:85	4	50	7696-12-0
		40:60	4	50	
11	,,	1R cis/trans 20:80	2	100	
		1R cis	1	100	51348-90-4
		1R trans	1	100	1166-46-7
12		(±) cis/trans	4	50	27223-49-0
13		nonchiral	1	100	24155-03-1
14		(±) cis/trans	4	50	23031-38-1
		1R cis/trans	2	100	

Common name	Code No.	Inventing institution	Basic producer* distributor	Trade names
	S 3488	Sumitomo		
		"		
Prallethrin	SF 4068 S 4068 SF	Sumitomo "	Sumitomo*	Pralle ETOC
Dimethrin		Boyce-Thomson		
Barthrin		Boyce-Thomson		
Tetramethrin		Sumitomo	Sumitomo*	Neopynamin (-D) Phthalthrin Synepirin
			Endura* Chimosa* Chinoin* Fairfield Amer. PR China*	
Futarthrin		Sumitomo	Sumitomo*	Neopynamin forte
Kikuthrin Proparthrin		Yoshitomi	Sumitomo	
	Y 4042	Yoshitomi		
Furamethrin Prothrin		Dainippon	Sumitomo Sumitomo	Pynamin D Pynamin-D-forte

Table 112. (continued)

Entry No.	Formula	Stereo-chemistry	No. of isomers	Content of active isomers %	CAS registry No.
15		(±) cis/trans	4	50	10453-86-8
16		(±) cis	2	50	10453-56-2
		(±) trans	2	50	10453-55-1
		1R trans	1	100	28434-01-7
17		1S trans	1	—	91224-53-2
		1R cis/trans 20:80	2	100	
18		1R cis	1	100	35764-59-1
19		1R trans	1	100	22431-62-5
20		1R cis-E	1	100	58769-20-3
21		1R cis-E	1	100	31192-70-8
22		(±) cis/trans	4	50	26002-80-2
23		1R cis/trans 20:80	2	100	26046-85-5
		1R cis	1	100	51186-88-0

Common name	Code No.	Inventing institution	Basic producer* distributor	Trade names
Resmethrin	NRDC 104	Rothamsted	Mitchell Cotts Fairfield American*	Pynosect Tetralat
	SBP		Pennick* Sumitomo*	Synthrin Benzofurolin Chryson
	FMC 17370		FMC*	Premgard Pyretherm
Isathrin cis-Resmethrin trans-Resmethrin				
Bioresmethrin	NRDC 107	Rothamsted		
			Wellcome Fisons	Resbuthrin
				Combat whitefly Biobenzofurolin
	RU 11484 FMC 18739 NIA 18739		Roussel Uclaf* FMC	
d-Resmethrin			Sumitomo*	Chryson forte
Cismethrin	NRDC 119 OMS 180 NIA 26021 RU 12063 FMC 26021	Rothamsted		
Bioethano-methrin	RU 11679	Roussel Uclaf	Roussel Uclaf*	K'Othrin
Kadethrin	RU 15525	Roussel Uclaf	Roussel Uclaf*	Kadethrin
	RU 12457	Roussel Uclaf		
Phenothrin	S 2539 OMS 1810	Sumitomo	Sumitomo	Welcide
d-Phenothrin	S 8100		Sumitomo*	Sumithrin
α-Phenothrin	S 2539 forte			Pesguard
Biophenothrin				Multicide

Table 112. (continued)

Entry No.	Formula	Stereo-chemistry	No. of isomers	Content of active isomers %	CAS registry No.
24		(±) cis/trans; RS	8	25	
		1R cis/trans; RS	4	50	
		20:80			39515-40-7
25		▲(±) cis/trans 40:60	4	50	52645-53-1
26		(±) cis	2	50	61949-76-6
					52341-33-0
		radioact.			63987-38-2
27		1R cis	1	100	54774-45-7
28		(±) trans	2	50	61949-77-6
					52918-63-5
		radioact.			52341-32-9
					61949-77-7
29		1R trans	1	100	51877-74-8
		1S trans	1	–	54774-46-8
30		1R cis	1	100	

Common name	Code No.	Inventing institution	Basic producer* distributor	Trade names
	S 2703	Sumitomo		
Cyphenothrin	S 2703 Aα		Sumitomo*	Gokilaht
	S 2703 forte RU 29209			
Permethrin	NRDC 143 SBP 1513	Rothamsted	Pennick*	Pounce, Pramec
	FMC 33297 PP 557		FMC* ICI*	Pounce Ambush. Kestral Ambushfog Ectiban; Kafil Perthrine; Picket
	WL 43479		Shell*	Talcord, Outflank Stockade
			Mitchel Cotts*	Permasect Pynosect
			Wellcome	Coopex; Perigen Stomoxin; Quamlin Nix
			Pan Britannica Ind. Sumitomo* Rhone Poulenc	Termite Eksmin; Adion Tornade; Corsair
	BW-21-Z			Priem Insectaban Atroban Overtime Permectin Hard-Hitter Guard Star Matadan Permanon
	NRDC 148			
	NRDC 167 NRDC 146			
Biopermethrin	NRDC 147			
		CSIR India		

Table 112. (continued)

Entry No.	Formula	Stereo-chemistry	No. of isomers	Content of active isomers %	CAS registry No.
31		(±) trans	2	50	
32		1R cis 1R trans	1 1	100 100	55700-98-6 55667-38-4
33		(±) cis/trans 40:60	8	18–25	52315-07-8
34		1R cis αS 1S cis αR	2	50	67975308 86753-92-6
35		1R trans αS	1	100	65732-07-2
36 37 38		1R cis αS (±) cis (±) trans	1 4 4	100 50 50	65731-84-2 66290-21-9 66290-20-8

Common name	Code No.	Inventing institution	Basic producer* distributor	Trade names
	FMC 55383	FMC		
	NRDC 157	Rothamsted		
	NRDC 163	Rothamsted		
Cypermethrin	NRDC 149	Rothamsted		
	PP 383		ICI*	Cymbush
				Imperator
				Kafil-Super
				Demon
				C-Methrin
				CCN 52
	WL 43467		Shell*	Ripcord, Stockade
	SH 1479			Cypofly, Barricade
				Folcord, Flectron
			FMC*	Ammo, Arrivo
			Mitchell Cotts*	Cyperkill
				Cypersect
				Afrisect
	S-035		Sumitomo*	Agrothrin
	EXP 5598		Rhone Poulenc	Sherpa, Shir
			Ciba Geigy	Polythrin
			Dow	Nurelle, Nural
			Farm Protection	Topple
	RU 24501		Roche-Maag	Mustrin
			Wellcome	Strike
	OMS 2002			
	SH 1479			Aradap
Alphamethrin	WL 85871	Shell	Shell*	Fastac, Fendona
				Fastact, Renegate
$\left(\begin{array}{c}\text{Cis-Cyper-}\\\text{methrin}\end{array}\right)$	SC 3027			Concord
	SH 5444		Halebank	Pouracide
				Ultimate
	SD 924549			Alpha
	FMC 39391		FMC*	Bestox
	FMC 65381			
			Ciba Geigy	Venom, Fenom
			Gharda* (India)	Bonsul; Dominex
	NRDC 182 $\left.\right\}$	Rothamsted		

Table 112. (continued)

Entry No.	Formula	Stereochemistry	No. of isomers	Content of active isomers %	CAS registry No.
39		1R cis αS	1	100	52918-63-5
40		1S cis αR	1	–	64346-01-8
		(±) cis/trans 40:60	8	20–25	68359-37-5
41		1R cisαS ⎫ 40 1S cis αR ⎭ 1R trans αS ⎫ 50 1S trans αS ⎭ } 4	4	50	
		(±) cis/trans; Z, E	16	6.5	69770-45-2
		(±) trans Z	4	25	
42		(±)	2	50	39515-41-8
43		(±) cis/trans 40:60	8	25	68523-18-2
44		(±) cis, Z	4	22	68085-85-8
45		1R cis Z, αS ⎫ 1R cis Z, αR ⎭	2	50	91465-08-6

Common name	Code No.	Inventing institution	Basic producer* distributor	Trade names
(Dekamethrin) Deltamethrin	NRDC 161			
	RU 22974		Roussel Uclaf*	Decis K'Othrin Slop, Butox Butoflin, Cislin
			Cooper Wellcome PR China*	Tackfly; Clout Slop; Crackdown
Cyfluthrin	BAY FCR 1272 OMS 2010	Bayer	Bayer*	Baythroid Solfac, Cylathrin Tempo Balecol Bayofly Responsar
$\left(\begin{array}{c}\text{Cyfluthrin K} \\ \text{+ L}\end{array}\right)$ Betacyfluthrin	Bay FCR 4545	Bayer		Bulldock
	BAY FCR 1622	Bayer		
Flumethrin	BAY FCR 2610 BAY VQ 1970	Bayer	Bayer*	Bayticol Drastic Deadline
Fenpropathrin (Fenpropanat)	S 3206	Sumitomo	Sumitomo*	Rody Danitol Meothrin Kilnmal
	DS 41706		Chevron Shell	Danitol Smash
Fenpyrithrin	Dow 417	Dow	Dow	Vivithrin
Cyhalothrin	PP 563 JF 289	ICI	ICI*	Grenade Heliocide Cyhalon Librecto Cypha
$\left[\begin{array}{l}\text{Cishalothrin} \\ \text{Cis-Cyhalothrin} \\ \text{Clocythrin} \\ \lambda\text{-Cyhalothrin}\end{array}\right]$ Lambda-cyhalothrin	PP 321	ICI	ICI*	Karate Matador ICON
			Nichino	

Table 112. (continued)

Entry No.	Formula	Stereo-chemistry	No. of isomers	Content of active isomers %	CAS registry No.
46		(±) cis/trans	8	25	60148-52-9
47		1R cis, R'S'; αS	2-	100	66841-25-6
48		1R cis, R'; αS (±) cis/trans	1 16	100 25	66841-26-7
49		(±) cis/trans	8	25	
50		1R cis E	1	100	
51		1R cis Z	1	100	
51a		1R cis Z	1	100	
52		(±) cis/trans	8	25	
53		(±) cis/trans	8	25	75528-07-2
54		(±)	4	20–25	51630-58-1

Common name	Code No.	Inventing institution	Basic producer* distributor	Trade names
Cypothrin	AC 206797	Amer. Cyanamid	Takeda Ishihara Nihon Noyaku Am. Cyanamid*	Cypothrin Panecto
Tralomethrin	RU 25474 HAG 107 NU 831 RU 24784	Roussel-Uclaf	Roussel Uclaf* Hoechst Sankyo	Scout D-End
Tralocythrin	CGA 74055	Ciba Geigy		
	DCH KKT 223	Kuraray		
	RU 40394	Roussel Uclaf		
	RU 39568 HR 187	,,		
	RU 38702	,,		
	NCI 85183 S 153	Nissan		
	NCI 85193 T 193	Nissan		
Fenvalerate	S 5602	Sumitomo	Sumitomo*	Sumicidin Sumifly Sumitic
	SD 43775		Shell*	Bellmark

Table 112. (continued)

Entry No.	Formula	Stereo-chemistry	No. of isomers	Content of active isomers %	CAS registry No.
55		(±) S; αS R; αR	2	50	67890-40-8
56		(+) S; αS	1	100	66230-04-4
57		(+) S; αRS	2	50	70124-77-5
58		RS; αRS S; αRs	4 2	25 50	69409-99-0 69409-94-5
59		S, αRs	2	50	76834-30-5
60		± (−) S	2 1	50 100	
61		(±)	4	~25	63935-38-6

Common name	Code No.	Inventing institution	Basic producer* distributor	Trade names
				Pydrin
				Sanmarton
				Ectrine
				Fenvale
				Y-Rich
				Acamethrin
				Pyrid, Agrofen
			United Phosphorus* (Ind)	
Fenvalerate A (Fenvalerate Aβ)	SK 102	Sumitomo	Sumitomo*	
	SH 43775		Shell*	
(Fenvalerate Aα) Esfenvalerate	SK 101 S 5602 Aα S 1844	Sumitomo	Sumitomo*	Sumialpha Sumi Gold
	SH 70616 MO 70616		Shell* Du Pont	Asana XL
$\left(\begin{array}{c}\text{"cis"-}\\ \text{Fenvalerate}\end{array}\right)$				
Flu-cythrinate	ACC 222705	Am. Cyanamid	Am. Cyanamide* Du Pont	Payoff; Cythrin Cybold Guardian
			Takeda Nissan	Fuching Jujr
	OMS 2007 A 1329311			
	ZR 3209	Zoecon		
Fluvalinate	ZR 3210	Zoecon	Zoecon/Sandoz	Mavrik HR Spur
			Hooker	
	MK 128		Mitsubishi	Klartan
	ZR 3902 SD 57706			
	GH 380 GH 401	CSIRO Austr. „		
Cycloprothrin Fencyclate	GH 414 NNI 802 NK 8116 NK 812	„	May & Baker Nihon Kayaku Roche Maag Du Pont/Dunlena	Cyclosal Baclash

Table 112. (continued)

Entry No.	Formula	Stereo-chemistry	No. of isomers	Content of active isomers %	CAS registry No.
62		(±)	4	25	78011-11-7
63		(±)	4	25	
64		(±) cis, Z	2	50	54774-45-7
		(±) trans Z	2	50	51877-74-8
65		1R cis, Z	1	100	92880-79-0
66		(±) cis Z, RS	4	25	93739-38-5 78383-23-0
67		(±) cis, Z	2	50	
68		(±) cis	2	50	
69		(±) cis/trans, Z	4	50	
70		(±) cis/trans RS, Z, E	16	25	54407-61-3
71		20:80 1R cis/trans RS, E	4	50	96895-17-9

Common name	Code No.	Inventing institution	Basic producer* distributor	Trade names
		Sumitomo		
		CSIRO Aust.		
Bifenthrin	FMC 4966	FMC	FMC*	Talstar
	FMC 55383			Brigade, Brookade Command Capture
Biphenate (Bufenthrin)	FMC 54800			
			Nissan Kanesho Rhone Poulenc	
	FMC 54617	FMC		
	FMC 51785	FMC		
	FMC 49655	FMC		
Butethrin		Taisho		
	S 3243	Sumitomo		
Empendrin	S 2852 forte	Sumitomo	Sumitomo*	Vaporthrin

Table 112. (continued)

Entry No.	Formula	Stereo-chemistry	No. of isomers	Content of active isomers %	CAS registry No.
72		1R trans	1	100	75867-00-4 67640-14-6
		1R cis	1	100	
73		(±) cis Z	2	50	76437-51-9
74		E	1	100	
75		nonchiral	1	100	80844-07-1
76		nonchiral	1	100	89764-44-3
77		nonchiral trans	1	100	
78		1R trans	1	100	118712-89-3
79		nonchiral		100	99503-10-3
80		nonchiral		100	105024-66-6
81		(±)	2	50	107796-06-5
81a		1R cis 20 1R trans 80	2	100	70062-52-1 72963-72-5
82		1R cis Z	1	100	

Common name	Code No.	Inventing institution	Basic producer* distributor	Trade names
Fenfluthrin	BAY NAK 1654 NAK 1901	Bayer Bayer		Remedor
Tefluthrin	PP 993	ICI	ICI*	Force
	FMC 60671	FMC		
Ethofenprox (Ethoxy- proxyfen)	MTI 500 MT 115000	Mitsui Toatsu	Mitsui Toatsu*	Trebon, Torebon Renatop Sparen
	MTI 800	Mitsui Toatsu		
	NRDC 200	Rothamsted		
Benfluthrin	NAK 4455	Bayer	Bayer*	
	SSI 116	Shionogi Sumitomo		
	HOE 498	Dainippon		
		ICI		
		Sumitomo		
	RU 39130	Roussel Uclaf		

Abbreviations

AU	australien	patent application	
BE	belgian	"	
BR	brazilian	"	
C.A.	chemical abstracts		
DOS	deutsche	Offenlegungsschrift	(german pat. appl.)
DAS	deutsche	Auslegeschrift	(german Patent
EP	european	pat. appl.	
FR	french	"	
GB	british	"	
HU	hungarian	"	
JA	japanese	"	
IL	israel	"	
NE	dutch	"	
US	United States	patent	
WO	world	pat. appl.	

Literature

1. Pyrethrum Post 7, 3 (1964).
2. Dainippon: JA 9597 (1956).
3. Pyrethrum Post 7, 41 (1964).
4. Mitchell, W.: Chem. and Ind. *1960*, 356.
5. FMC: US 3087854 (1962).
6. Levy, L. W.: US 3083136 (1962).
7. Mitchell Cotts Pyrh. Ltd.: GB 1031688 (1963).
8. Pennick: US 3042706 (1960).
9. Dainippon: JA 351044630 (1974)
10. Staudinger, H., Ruzicka, L.: Helv. Chim. Acta *1924*, 201.
11. Yamamoto, R.: J. Chem. Soc. Jap. *44*, 311 (1923).
12. LaForge, F. B., Soloway, S. B.: J. Am. Chem. Soc. *69*, 2932 (1947).
13. Sawicki, R. M., Thain, E. M.: J. Sci. Food Agric. *13*, 202 (1962).
14. Elliott, M.: Environ. Health Perspectives *14*, 3 (1976).
15. Sawicki, R. M.: J. Sci. Food Agric. *13*, 172 (1962).
16. Crombie, L., Harper, S. H.: J. Chem. Soc. *1954*, 470.
17. This pictorial nomenclature proposed by Elliott (52) is much more convenient for chemists.
18. Begley, M. J., Crombie, L., Simonds, D. J., Whiting, D. A.: J. Chem. Soc. Chem. Commun. *1972*, 1276.
19. Wickham, J. C.: Pestic. Sci. *1976*, 273.
20. Pattenden, G., Store, R.: Tetrahedron Lett. *1973*, 3473.
21. Tamelen v., E. E., Schwartz, M. A.: J. Am. Chem. Soc. *93*, 1780 (1971).
22. Godin, P. J.: J. Chem. Soc. *1963*, 5878.
23. Epstein, W. W., Gaudioso, L. A.: J. Org. Chem. *44*, 3113 (1979).
24. Alexander, K., Epstein, W. W.: J. Org. Chem. *40*, 2576 (1975).
25. LaForge, R. B.: J. Org. Chem. *2* (1937) 309.
26. Chem. Ind. (Düsseldorf) 3087 (1978).
27. Naturwissenschaftliche Rundschau *33* (1980) 404.
28. Kueh, S. H. J., MacKenzie, J. A., Pattenden, G.: Plant cell Rep. *1985*, 118.
29. Zieg, R. G., Zito, S. W., Staba, E. J.: Planta Med. *1983*, 88.
30. McLaughlin, Gormley King: EP 124049 29. 4. 83/7. 11. 84.
31. Univ. of Minnesota: US 4525455 29. 4. 83/25. 6. 85.
32. Farm Chemicals *1981*, 10, 40.
33. Staba, E. J., Nygard, B. G., Zito, S. W.: Plant Cell Tiss. Org. Cult. *3* (1984) 211.
34. Pandita, P. N., Bhat, P. K.: Herba Hung. *23* (1984) 89.
35. Bestmann, H. J., Claassen, B., Kobold, U., Vostrowsky, O., Klingauf, F., Strobel, H., Knobloch, K.: Z. Naturforsch. C. Biosci *1984*, 543.
36. Hollrung, M.: Die Mittel zur Bekämpfung der Pflanzenkrankheiten, Berlin 1914, p. 37.
37. Pyrethrum, The Natural Insecticide, Ed. Casida, J. E. by Academic Press 1972.
38. Otieno, D. A., in: Natural Products for Innovative Pest Management, Ed. Whitehead and Bowers, Pergamon Press 1983.
39. Elliott, M., Briggs, J.: Pestic. Sci. *1976*, 326.
40. Hung Hee Lee, L.: Pestic. Sci. *1976*, 258.

41. Matsu, M., Nishimura, H., Fujita, T.: Pestic. Biochem. Physiol. 25 (1986) 288.
42. Staudinger, H., Ruzicka, L.: Helv. Chim. Acta 1924, 177, 201, 212, 236, 245, 390, 406.
42a. Yamamoto et. al.: J. Tokyo Chem. Soc. 80, 126 (1919); J. Chem. Soc. Jap. 44, 311 (1923); 44, 1070 (1923).
43. Staudinger, H., Ruzicka, L.: ibid. 1924, 456.
44. Harville, E. K.: Contrib. Boyce-Thomson-Inst. 10, 143 (1939).
45. Schechter, M. S., LaForge, R. B., Greene, N.: J. Am. Chem. Soc. 71, 3165 (1949).
46. Barthel, W. F.: Advance Pest. Controll Res. 4 (1961) 33.
47. Boyce-Thomson-Inst.: US 248658 (1949).
48. Boyce-Thomson-Inst.: US 2886485 (1957).
49. Farkas, J., Sorm, F., Kourim, P.: Col. Czech. Chem. Commun. 24, 2230 (1959).
50. Kato, T., Ueda, K., Fujimoto, K.: Agric. Biol. Chem. 28 (1964) 914.
51. Elliott, M., Farnham, A. W., Janes, N. F., Needham, P. H., Pearson, B. C.: Nature 213 (1967) 493.
52. Elliott, M., Farnham, A. W., Janes, N. F., Needham, D. H., Pulman, D. A.: Nature 248 (1974) 710.
53. Ohno, N., Fujimoto, K., Okuno, Y., Mizutani, T., Hirano, M., Itaya, N., Honda, T.: Agric. Biol. Chem. 38 (1974) 881.
54. Fujimoto, K., Itaya, N., Okuno, Y., Kadoka, T., Yamaguchi, T.: Agric. Biol. Chem. 37 (1973) 2681.
55. Matsuo, T., Itaya, N., Mizutani, T., Ohno, N., Fujimoto, K., Okuno, Y., Yoshioka, H.: Agric. Biol. Chem. 40 (1976) 247.
56. Holan, G. B., O'Keefe, D. F., Walser, R. A., Virgona, C. T.: Nature 272 (1978) 734.
57. Behrenz, W., Naumann, K.: Pflanzenschutz-Nachrichten Bayer 35 (1982) 309.
58. Hamann, J., Fuchs, R.: Pflanzenschutz-Nachrichten Bayer 34 (1981) 121.
58b. Entley, P. B., Cheetham, R., Huff, R. K., Swanborough, J.: Pestic. Sci. 1980, 158.
59. Stendel, W., Fuchs, R.: Vet. Med. Rev. 1982, 115.
60. Henrick, C. A., Garcia, D. A., Staal, G. D., Cerf, D. C., Anderson, R. J., Gill, K., Chinn, H. R., Labowitz, J. N., Leippe, M. M., Woo, S. L., Carney, R. L., Gordon, D. C., Krohn, G. K.: Pestic, Sci. 1980, 224.
61. Bull, M. J., Davies, J. H., Searle, R. J. G., Henry, A. C.: Pestic. Sci. 1980, 249.
62. Udagawa, T., Numato, S., Oda, K., Shiraishi, S., Kodaka, K., Nakatami, N.: Proc. Recent. Adv. Chem. Insect. Controll, Symposion Cambridge Sept. 1985, 192.
63. PCT: WO 822321 (1980).
64. Clemens, A. N., May, T. W.: Pestic. Sci. 1977, 661.
65. Nishimura, K., Kobayashi, T., Fujita, T.: Pestic. Biochem. Physiol. 25 (1986) 387.
66. Holan, G.: Recent Advances in the Chemistry of Insect Controll, Proc. Sympos. Cambridge 1984, 114.
66a. Poppleton, B. J.: Acta Crystallogr. C Cryst. Struct. Commun. 1986, 879.
67. Holan, G.: Proc. Neurotox' 79, 73 Symposion York Sept. 1979.
68. Elliott, M., Janes, N. F.: Chem. Rev. 7 (1978) 473.
69. Hammann, J., Hartmann, A.: (Bayer, unpubl. 1979).
70. Scharf, H. D., Kalkhoff, K., Janus, J.: Tetrahedron 35 (1979) 2513.
71. Elliott, M., Janes, N. F.: Pestic. Sci. 1983, 189.
72. Behrens, W., Hammann, J., Naumann, K.: (Bayer, unpublished 1977).
73. Graham-Bryce, I. J.: Rothamsted Report 1976, I, 159.
74. Tessier, J. R.: Proc. Sympos. Cambridge Sept. 1984, Recent Advances in Chem. Insect Controll, 26, Ed. Janes, N. F., Royal Soc. Chem. London.
74a) Cora, D., Arnoldi, A., Rossini, L.: Arch. Toxicol. Suppl. 1986, 329.
76. Sugiyama, T., Kobayashi, A., Yamashita, K.: Agr. Biol. Chem. 38 (1974) 979.
77. Reid, J. R., Marmor, R. S.: J. Org. Chem. 43 (1978) 999.
78. Hammann, J., Fuchs, R.: (Bayer, unpublished 1980).
79. Brady, W. T., Norton, S. J.: J. Agric. Food Chem. 37 (1985) 1026.
80. Mack, H.: Diss. Univ. of Tübingen, 1985.
81. Hoechst: DOS 3044011 (1980).
82. Union Explosivos Rio Tinto: DOS 3445504 (1984).
83. Ogiermann, L., Nauk, Pr.: Inst. Ochr. Rosl. 25, 5 (1983), C. A. 103, 18351.

84. Kuraray: JA 83720 (1976).
85. FMC: EP 14593 (1979).
86. Sirrenberg, W.: (Bayer, unpubl. 1978).
87. Elliott, M., Elliott, R. L., Janes, N. F., Kambay, B. P. S., Pulman, P. A.: Pestic. Sci. *1986*, 691.
88. Elliott, M., Elliott, R. L., Janes, N. F., Kambay, B. P. S.: Pestic. Sci. *1986*, 701.
89. Elliott, M., Janes, N. F.: Pestic. Sci. *1982*, 413.
90. Henrick, C. A.: Recent Advances in the Chemistry of Insect Controll, 133 Proc. Symposion Cambridge 1984, Ed.: Janes, N. F., Royal Soc. Chem. London.
91. Matsuo, N., Jano, T., Yoshiska, H., Mori, K.: Pestic. Chem. Hum. Welfare Environ. Proc. IUPAC Congr. Kyoto (1982) I, 279.
92. Hammann, I., Fuchs, R., Stendel, W., Naumann, K.: Paper delivered by Naumann at the 190th ACS Meeting, Chikago Sept. 1985.
93. Holan, G., Johnson, W. M., Virgona, C. T., Walser, R. A.: J. Agric. Food Chem. *1986*, 520.
94. Elliott, M., Janes, N. F. et al.: Advances in Pestic. Sci., Proc., IUPAC Conf. Zürich 1978 II, 166.
95. Nakatani, K., Numata, T., Inoue, K., Oda, K., Kodaka, K., Udagawa, M., Gohbara, M., Shiraishi, S.: 190th ACS Meeting, Chikago, Sept. 1985.
96. Elliott, M., Janes, N. F. et al.: Phytiatrie-Phytopharmazie, 27, 99 (1978).
97. Kohn, G. K.: ACS Symposium Series 1974, Mechanism of Pesticide Action.
98. Elliott, M.: Comp. Rev. 7, 473 (1978).
99. Elliott, M., Janes, N. F. et al.: Pestic. Sci. 1974, 796.
100. Elliott, M. et al.: Pestic. Sci. *1975*, 537.
101. Nakayama, J., Ohno, N., Aketa, K., Suzuki, Y., Kato, T., Yoshioka, H.: Advances Pesticide Sci., Proc. IUPAC Conf. Zürich 1978, II, 174.
102. Ohno, N., Fujimoto, K. et al.: Pestic. Sci. 1976, 241.
103. Ohkada, K., Fujimoto, K., Okuno, Y., Matsui, M.: Agr. Biol. Chem. *37*, 2205 (1973).
104. Elliott, M. et al.: Pestic. Sci. *1978*, 112.
105. Elliott, M. et al.: Nature *244*, 456 (1973).
106. Nishimura, K., Narahashi, T.: Pestic. Biochem. Physiol. *8*, 53 (1978).
107. Tessier, J. R., Teche, A. P., Demoute, J. P.: Pestic. Chem. Human Welfare Environ. Proc. IUPAC Congr. Kyoto 1982, I, 95.
108. Elliott, M. et al.: Chem. Rev. *1978*, 473.
109. Elliott, M. et al.: Pestic. Sci. *1974*, 681.
110. Elliott, M. et al.: Pestic. Sci. *1980*, 120.
111. Bendley, P. D., Cheetham, R., Huff, R. K., Pascoe, R., Sayle, J. D.: Pestic. Sci. *1980*, 156.
112. Tessier, J. R.: Paper delivered at the 190th ACS Meeting, Chikago Sep. 1985.
113. Bosone, E., Corda, F., Gozo, F., Menconi, A., Piccardi, P., Caprioli, V.: Pestic. Sci. *1986*, 621.
114. Barlowe, F., Hadaway, A. B., Flurer, L. S., Grose, J. E., Turner, G. R.: Pestic. Sci. *1977*, 291.
115. Nolan, J., Roulston, W., Barton, R. H.: Pestic. Sci. *1977*, 484.
116. Soderlund, D. M.: Proc. Neurotox' *79*, 449, Symposion York Sept. 1979, 449.
117. Sumitomo: EP 56271 (1981).
118. Tsushima, K., Matsuo, N., Itaya, N., Yano, T., Hatakoshi, N.: Pestic. Chem. Human Welfare Environ. Proc. IUPAC Congr. Kyoto 1982, I, 91.
119. Aketa, K., Ohno, N., Itaya, N., Nakayamo, J.: Agr. Biol. Chem., *42*, 845, 1978.
120. Aketa, K., Ohno, N., Itaya, N., Nakayamo, J.: Paper delivered on the International Symposium on Chemistry of Pyrethroids, Oxford July 1979.
121. Wheeler, T. N.: J. Agric. Food Chem. *1984*, 1125.
122. Henrick, C. A., Anderson, R. J., Staal, G. B.: Pestic. Chem. Human Welfare Environ. Proc. IUPAC Congr. Kyoto 1982 I, 107.
123. Ayad, H. M., Wheeler, T. N.: J. Agric. Food Chem. *1984*, 85.
124. Holan, G.: Nature, *272*, 73 (1978).
125. Holan, G., Johnson, W. M. P., O'Keefe, D. F., Rihs, K., Smith, D. R. J., Virgona, C. T., Walser, R., Haslam, J. M.: Pestic. Chem. Human Welfare Environ. Proc. IUPAC Congr. Kyoto 1982 I, 1919.
126. Janes, N. F.: 190th ACS Meeting, Chikago Sept. 1985.

126a. Bently, P. D., Cheetham, R., Huff, R. K., Pascoe, R., Sayle, J. D.: Pestic. Sci. *1980*, 156.
127. Anderson, R. J., Adams, K. G., Henrick, C. A.: J. Agric. Food Chem. *33*, 508 1985.
128. Nishimura, K., Kobayashi, T., Fujita, T.: Pestic. Biochem. Physiol. *25*, 387, 1986.
129. Gewolt, P.: Pestic. Sci. *1976*, 604.
130. Breese, M. H.: Pestic. Sci. *1977*, 264.
131. Bull, M. J., Davies, J. H., Searle, R. J. G., Henry, A. C.: Pestic. Sci. *1980*, 249.
132. Sumitomo: Technical Information *1981–1984*.
133. Kohn, G. M.: ACS Symp. Ser. *1974*, Mechanisms of Pesticid Action.
134. Elliott, M. et al.: Pestic. Sci. *1971*, 115.
135. Ford, M. G.: Pestic. Sci. *1979*, 39.
136. Elliott, M. et al.: J. Sci. Food Agric. *18* (1967) 325.
137. Ohno, N., Fujimoto, K. et al.: Agr. Biol. Chem. *40*, 247, (1976).
138. Hirano, M.: Jap. J. Sanit. Zool. *29*, 219, 1978.
139. Elliott, M., Janes, N. F., Kambay, B. P. S., Pulman, D. A.: Pestic. Sci. *1983*, 182.
140. Elliott, M. et al.: Pestic. Sci. *1982*, 407.
141. Hirano, M., Yano, T., Matsuo, N., Kitamura, S., Nishioka, T., Fujita, Y.: Eisei Dobutsu *1983*, 263.
142. Elliott, M., Farnham, A. W., Janes, N. F., Johnson, D. M., Pulman, D. A.: Pestic. Sci. *1980*, 513.
143. ICI Austr.: EP 6354 (1978).
144. Malhotra, S. K., Heertum, C., Larsson, L. L., Ricks, M. J.: J. Agric. Food Chem. *29*, 1287 (1981).
145. Fuchs, R., Klauke, E., Hammann, J.: (Bayer, unpubl. 1980).
146. Plummer, E. L., Pincus, D. S.: J. Agric. Food Chem. *1981*, 1118.
147. Plummer, E. L., Stewart, R. S.: J. Agric. Food Chem. *1984*, 1116.
148. Elliott, M.: Chemistry and Ind. *1969*, 779.
149. Hammann, I., Behrenz, W., Naumann, K.: Paper delivered at the 190th ACS-Meeting, Chikago Sept. 1985.
150. McDonald, E., Punja, N., Jutsum, A. R.: Brit. Crop. Prot. Conf. *1986*, 199.
151. Ando, T., Ruzo, L. O., Engel, J. L., Casida, J. E.: J. Agric. Food Chem. *1983*, 250.
151a. Elliott, M.: Proc. Recent Advances in Insect Controll, Symposion Cambridge Sept. 1984, Ed.: Janes, N. F.: Royal Soc. Chem. London p. 73.
152. Chinoin: DOS 2906928 (1978).
153. Sumitomo: EP 50544 (1980).
154. Elliott, M., Farnham, A. W., Janes, N. F., Kambay, B. P. S.: Pestic. Sci. *1981*, 503.
156. Donner, W., Hammann, I., Naumann, K.: (Bayer unpublished 1979). No other physical data did correlate with the biological data.
157. The corresponding chrysanthemates were not active. Tsu, H. T., Brady, T., Norton, S. J.: J. Agric. Food Chem. *1985*, 751.
158. Yura, Y.: Agr. Biol. Chem. *42*, 1767 (1987).
159. Plummer, E. L., Sciders, R. P., Seely, D. E., Stewart, R. R.: Pestic. Sci. *1984*, 509.
160. Engel, J. F., Staetz, C. A., Young, S. T., Crosby, G. A.: Proc. Recent Advances in Insect Controll, Symposion Cambridge Sept. 1984, Ed.: Janes, N. F., Royal Soc. Chem. London, p. 162.
161. Nakada, Y., Yura, A., Ohno, S., Yoshimoto, M.: Agric. Biol. Chem. *42*, 1357 (1978); *42*, 1365 (1978); *42*, 1767 (1978).
162. Mobil Oil: EP 4754 (1978).
163. Shell: US 4434182 (1978).
164. Nanjyo, K., Katsuyama, N., Yamamura, T., Hyeou, S. B., Suzuki, A., Tamura, S.: Agric. Biol. Chem. *44*, 218 (1980).
164a. Svendson, A., Pedersen, L. E., Klemmensen, P. D.: Pestic. Sci. *1986*, 93.
165. Elliott, M.: 190th ACS Meeting, Chikago Sept. 1985.
166. Holan, G., Johnson, W. M. P., Jarvis, K. E., Virgona, C. T., Walser, R. A.: Pestic. Sci. *1986*, 715.
167. Sumitomo: EP 175377 (1984).
168. Mitsui Toatsu: DOS 3402483 (1983).
169. Casida, J. E., Kirino, O.: J. Agric. Food Chem. *33*, 1208 (1985).

169a. Casida, J. E. et al.: J. Agric. Food Chem. *5* 931 (1969).
170. Hirano, M., Itaya, N., Ohno, J., Fujita, J., Yoshioka, H.: Pestic. Sci. *1979*, 291.
171. Leake, L. D., Lauckner, S. M., Ford, M. G.: Proc. Neurotox' 79, 423, Symposion York Sept. 1979.
172. Sumitomo: DOS 3028290 (1980).
173. Dainippon: US 4459305 (1980).
174. Fuchs, R., Hammann, I., Naumann, K.: (Bayer unpublished 1979).
175. Sugiyama, T., Kobayashi, A., Yamashita, K.: Agric. Bull. Chem. *38*, 979 (1974).
176. Martell, J. J.: Pestic. Chem. Human Welfare Environ. Proc. IUPAC Congr. Kyoto 1982, II, 165.
177. Miyakado, M. et al.: Agr. Biol. Chem. *39*, 267 (1975).
178. Begley, M. J. et al.: J. Chem. Soc. chem. Com. *1972*, 1276.
179. Matsuo, N., Yano, T., Yoshioka, H., Kuwahara, S., Mori, K.: Agric. Biol. Chem. *45*, 1915 (1981).
179b. Sumitomo: DOS 2356702 (1972).
180. Grünig, R., Pospischil, R., Cymorek, S., Metzner, W.: 7th Ann. Meeting Inter. Res. Group on Wood Preservation, France May 1986.
180a. Ford, M. G.: Pestic. Sci. *1979*, 43.
181. ICI: EP 31199 (1979).
181a. DOS 2356706 (1972).
182. Owen, J. P.: J. Chem. Soc. P.T. I, *1975*, 1865.
183. Owen, J. P.: Acta Cryst. *37*, 1311 (1981).
184. Owen, J. P.: J. Chem. Soc. P.T. I *1975*, 1233.
185. Born, L.: (Bayer unpublished 1984).
186. Heritage, K. J.: Biochemical Soc. Trans. *1982*, 310.
187. Hopfinger, A. J., Malhotra, D., Battershell, R. D., Ho, A. W.: J. Pestic. Sci. *9*, 631 (1984).
188. Hopfinger, A. J., Battershell, R. D.: Adv. Pestic. Sci. Proc. IUPAC Congr. Zürich 1978 II, 196.
189. Plummer, E. L.: 190th ACS Meeting, Chikago Sept. 85.
      Plummer, E. L.: ACS Symp. Ser. *255*, (1984) 297.
190. Lee, H. H.: Pestic. Sci. *1976*, 258.
191. Ford, M. G., Greenwood, R., Leake, L. D., Szydlo, R. M., Turner, C. H.: Pestic. Sci. *16*, 673 (1985).
192. Jones, O. T., Lee, A. G.: Pestic. Biochem. Physiol. *1986*, 431.
193. Chang, C. P., Plapp, F. W.: J. Econ. Entomol. *76*, 1206 (1983).
194. Holan, G. et al.: Pestic. Sci. *1984*, 632.
195. Narahashi, T.: Comp. Biochem. Physiol. *72*, 411 (1982).
196. Nakagawa, S., Okajima, N., Nishimura, K., Fujita, T., Nakayima, N.: Pestic. Biochem. Physiol. *17*, 259 (1982).
197. Fujita, T.: Adv. Pestics. Sci., IUPAC Congr., Zürich 1978, Abstr. II-25.
198. Plummer, E. L., Pinkus, D. S.: J. Agric. Food Chem. *29*, 1121 (1981).
199. Plummer, E. L.: J. Agric. Food Chem. *31*, 718 (1983).
200. Hatakoshi, M., Takayama, C., Matsuo, N., Nakayama, J., Kirino, O.: Nippon Noyaku Gakkaishi *1983*, 179.
201. Ford, M. G., Greenwood, R.: Paper delivered at the Neurotox' 85 Symposion Bath April 1985.
202. Szydlo, R. M., Ford, M. G., Greenwood, R., Salt, D. W.: Pharmacochem. Libr. *1983*, 203.
203. Szydlo, R. M., Ford, M. G., Greenwood, R., Salt, D. W.: QSAR Des. Bioact. Compd. *1984*, 219–237.
204. Szydlo, R. M., Ford, M. G., Greenwood, R., Salt, D. W.: QSAR Des. Bioact. Compd. *1984*, 301.
205. Ruzo, L. O.: Biochem. Physiol. *15*, 137 (1981).
206. Gammon, D. W.: Proc. Brit. Crop. Conf. Pest. Div. *1984*, 871.
      Gammon, D. W., Ruzo, L. O.: Pestic. Sci. *15*, 586 (1984).
207. Irving, S. N.: Proc. Brit. Crop. Conf. Pest. Div. *1984*, 859.
207a. Kaneko, H., Takamatsu, Y., Kitamura, N., Yoshitake, A., Miyamoto, J.: Nippon Noyaku Gakkaishi *11*, 533 (1986).

208. Elliott, M., Janes, N. F.: Proc. Brit. Crop. Conf. *1984*, 849.
209. Benoit, M., Roche, M., Carla, P.: Abstr. Neurotox' *85* 13, Symposion Bath April 1985.
210. Meinhard, C., Bruneau, P., Roche, M.: J. Chromatogr. *1985*, 349.
211. Roche, M., Frelin, C., Bruneau, P., Meinard, C.: Pestic. Biochem. Physiol. *24*, 306 (1985).
211a. Yamanoi, F.: Nippon Sanshigaten Zasshi *55*, 329 (1986).
212. Soderlund, D. M.: Pestic. Chem. Human Welfare Environ. Proc. IUPAC Congr. Kyoto 1982 III, 69.
213. Pichon, Y., Guillet, J. C., Pelhate, M.: Pestic. Sci. *1985*, 627.
213a. Behrenz, W.: (Bayer, unpubl. 1980).
214. Bowers, W. S., Haunerland, N. H.: Arch. Insect. Biochem. Physiol. *1986*, 87.
215. Gerolt, Ph.: Biol. Rev. Cambridge Philos. Soc. *58*, 223 (1983); Pestic. Sci. *1975*, 233.
216. Madrell, S. H. P.: Proc. Neurotox' *79*, 329, Symposion York Sept. 1979.
217. Shipp, E.: Proc. Neurotox' *79*, 441, Symposion York Sept. 1979.
218. McDonald, S.: J. Econ. Entomol. *74*, 46 (1981).
219. Whitney, W. V.: Proc. Brit. Crop. Conf. *1979*, 387.
219a. Subramanyam, Bh., Cutkomp. L. K.: Exper. Appl. Acarology *3*, 109 (1987).
220. Gammon, D.: Pestic. Sci. *1978*, 79.
221. Scott, J. G., Matsumura, F.: Pestic. Biochem. Physiol. *19*, 141 (1983).
222. Sparks, T. C., Parloff, A. M., Rose, R. L.: J. Econ. Entomol. *76*, 243 (1983).
222a. Brown, M. A.: J. Econ. Entomol. *80*, 330 (1987).
223. Burgess, L., Hinks, C. F.: Can. Entomol. *118*, 79 (1986).
224. Pietkiewicz, J., Pawinska, M.: Mater. Ses. Nauk. Inst. Ochr. Rosl. *25*, 451 (1985); C.A. *105*, 204738.
225. Grafins, E.: J. Econ. Entomol. *79*, 588 (1986).
226. Fuchs, R., Hammann, I.: Pflanzenschutznachr. Bayer *34*, 121 (1981).
226a. Fuchs, R., Homeyer, B.: (Bayer, unpublished 1979).
226b. Soderlund, D. M., Hessney, C. W., Hellmuth, D. W.: Pestic. Biochem. Physiol. *20*, (1983) 161.
227. Ishaaya, J., Casida, J. E.: Pestic. Chem. Human Welfare Environ. Proc. IUPAC Congr. Kyoto 1982 III 307.
228. Soderlund, D. M., Sanborn, J. R., Lee, P. W.: Progress in Pestic. Biochem. III 401 Ed. Hutson, Roberts; Wiley 1983.
229. Nakagawa, S., Nishimura, K., Kurihara, N., Fujita, T.: Pestic. Biochem. Physiol. *24* (1985) 182.
230. Casida, J. E.: Pestic. Chem. Human Welfare Environ. Proc. IUPAC Congr. Kyoto 1982, III, 307.
231. Soderlund, D. M.: Abstr. ACS Meeting 1981, 13.
232. Ishaaya, J., Casida, J. E., Ascher, K. R. S., Elsner, A.: Pestic. Sci. *14*, 367 (1983).
233. Ford, M. G., Szydlo, R. M., Salt, D. W.: Proc. Brit. Crop. Prot. Conf. *1984*, 865.
234. Soderlund, D. M.: J. Envir. Sci. Health, B 1983, *318*, 15.
235. Behrenz, W., Elbert, A.: Anzeiger für Schädlingskunde etc. *58*, 30 (1985).
236. v. d. Bercken, J., Vijverberg, H.: Paper delivered at the Neurotox' *85*, Symposion Barth April 1985.
    Buckley, D. S., Leake, L. D., Ford, M. G.: Pestic. Sci. *1985*, 545.
237. Ruigt, G. S. F., Klis, J. F. L., v. d. Bercken, J.: J. Comp. Physiol. *1986*, 43.
237a. Rajavel, A. R., Vasuki, V., Paily, K. P., Ramiah, K. D., Mariappan, T., Kalyanasundaraman, M., Tyagi, B. K., Das, P. K.: Indian J. Med. Res. *1987*, 168.
238. Mulla, M. S.: Proc. Ann. Conf. Calif. Mosquito Vector Control Assoz. *1980*, 92.
239. Quinlan, R. J., Gatehouse, A. G.: Pestic. Sci. *1981*, 439.
239a. Molchanov, M. J., Kuteev, F. S., Molchanova, Y. A., Kotovskaya, A. P.: Prikl. Biokhim. Mikrobiol. *23*, 253 (1987).
240. v. d. Bercken, J.: Priv. Communication.
240a. Blomquist, J. R., Miller, T. A.: Arch. Insect. Biochem. Physiol. *1986*, 551.
241. Scott, J. G., Georghiu, G. P.: Pestic. Biochem. Physiol. *21*, 53 (1984).
242. Hirano, M. (Sumitomo): Paper presented at Bayer AG-Monheim, German Phyto-Medicinal Society, 14. 3. 85.

243. Tan, K. H.: Pestic Sci. *1981*, 619.
244. Armstrong, K. F., Bonner, A. B.: Pestic. Sci. *16*, 641 (1985).
245. Wong, R. Y.: US Pat. 4357336.
246. Shemanchuk, J. A., Taylor, W. G.: Pestic. Sci. *15*, 557 (1984).
248. Schreck, C. E., Snoddy, E. L., Mount, G. A.: J. Econ. Entomol. *73* (1980) 436.
     Hartz Mountain Corp: US 4547360 (1983).
248a. Bueschert, M. D., Rutledge, L. C., Wirtz, R. A.: Pestic. Sci. *1987*, 165.
249. Harvey, T. L., Brethour, J. R.: J. Econ. Entomol. *72*, 532 (1979).
250. Kahn, M. A., Colwell, D. D.: Pestic. Sci. *15*, 487 (1984).
250a. Earth Seigaku: JA 5692803 (1979).
251. Haynes, K. F., Baker, T. C.: Arch. Insect. Biochem. *2*, 283 (1985).
251a. Maddison, P. A.: J. Econ. Entomol. *66* (1973) 591.
252. Gist, G. L., Pless, C. D.: Fla. Entomol. *68*, 450 (1985).
253. Jackson, A. E. A., Richard, R. M.: Pestic. Sci. *1985*, 364.
254. Injac, M., Dulic, K.: Proc. Brit. Crop. Prot. Conf. Pest. Div. *3*, 1171 (1984).
255. Ho, S. H., Go, P. M.: Toxicol. Lett. *22*, 161 (1984).
256. Khodzhaev, Sh. T., Eshmatov, O. T.: Zashch. Rast. *7*, 24 (1983).
257. Gist, G. L., Pless, C. D.: Fla. Entomol. *68*, 462 (1985).
258. Bayer: DOS 2840992 (1977).
259. ICI: US 4370346 (1983); US 4405640 (1983).
260. McDonald, E., Punja, N.: Pestic. Sci. *17*, 459 (1986).
260a. Mc.Donald, E., Punja, N., Jutsum, A. R.: Proc. Brit. Crop. Prot. Conf. *1986*, 97; 199.
261. Staetz: US Def. Publ. T 107001 22. 8. 84/2. 9. 86.
262. FMC: WO 85/4553 13. 4. 84/24. 10. 85.
     (The 2-isomers are claimed to be more acaricidally active).
263. French-Constant, R. H., Vickerman, G. P.: Entomophaga *30*, 271 (1985).
264. Roussel Uclaf: US 4181735 (1977).
265. Inglesfield, C.: Bull. Environ. Contam. Toxicol. *33*, 568 (1984).
266. Lhoste, L., Hotellièr, M. Deltamethrin Monograph, Roussel Uclaf 1982 (9).
267. Bayer Pflanzenschutzbrief *B*, 115 (1980).
268. Hull, L. A., Starner, V. R.: J. Econ. Entomol. *76*, 122 (1983).
269. Iftner, D. C., Hall, F. R.: J. Agric. Entomol. *1*, 191 (1984).
270. McKee, M. J., Knowles, C. O.: J. Econ. Entomol. *77*, 1376 (1984).
271. Cranham, J. E., Easterbrook. M. A., Kapetanakis, E.: Pestic. Sci. *1985*, 214.
271a. Heath, J.: Meded. Fac. Landbouwwet. Rijksuniv. Gent *50* (1985) 665.
272. Chelliah, S., Heinrichs, E. A.: Environ. Entomol. *9*, 773 (1980).
273. Penman, D. R., Chapman, R. B., Bowie, M. H.: Proc. N.Z. Weed Pest. Control Conf. *1984*,
     262.
274. FMC: US 4552892 (1983).
275. Osawa, K.: Pestic. Chem. Human Welfare Environ. Proc. IUPAC Congr. Kyoto 1982 II,
     91.
276. Roussel Uclaf: EP 93664 (1982).
277. Martell, J. J.: 190th ACS Meeting Chikago Sept. 1985.
278. Shell: EP 2091 (1977).
279. Shell: EP 229 (1977).
280. Roussel Uclaf: DOS 3131912 (1980).
281. Stendel, W., Fuchs, R.: Acarol. *2*, 1252 (1984).
282. Hamel, H. D., Dorn, H.: Acarol. *2*, 1263 (1984).
282a. Bayer: DOS 3529693 (1985).
283. Stendel, W., Fuchs, R.: Vet. med Rev. *1982*, 115.
285. Stubbs, V. K., Wilshire, C., Webber, L. G.: Aust. Vet. J. *59*, 152 (1982).
286. Earth Chemical Co.: JA 604231 (1983).
287. Roussel Uclaf: DOS 2707119 (1976).
     Applipharm.: FR 2555449 (1983).
288. Schreck, C. E., Shoddy, E. L., Mount, G. A.: J. Econ. Entomol. *73*, 436 (1980).
289. Georghiu, G. P.: Resistance Documentation Centre Univ. of California Riverside, July
     1980.

290. Keiding, J. P.: Pestic. Sci. 7, 283 (1976).
291. Sawicki, R. M.: Progr. Pestic. Biochem. Toxicol. 1985, 143.
292. Alekseev, A. N.: Med. Parazitol. Parazit Bolezni 51, 28 (1982).
293. Malinowski, H.: Pol. Pismo Entomol. 50, 559 (1980).
294. Prasittisuk, Ch., Bosvine, J. R.: Pestic. Sci. 1977, 527.
295. Chapin, J., Wasserstrom, R.: Nature 293, 181 (1981).
296. Halliday, W. R., Georghiu, G. P.: J. Econ. Entomol. 78, 1227 (1985).
297. Gunning, R. V., Easton, C. S., Greenup, L. R., Edge, V. E.: J. Econ. Entomol. 77, 1283 (1984).
298. Daly, J. C., McKenzie, J. A.: Proc. Brit. Crop. Prot. Conf. 1986, 951.
299. El. Sayed, E. I., Abodonia, S. A., Homanna, A. H.: East Afric. Agric. For J. 1980/81 46, 13 (1985).
300. Heather, D. W.: J. Stored Product Res. 22, 15 (1986).
301. Kunz, J. E., Schmid, C. D.: J. Agric. Entomol. 2, 358 (1985).
302. Sheppard, D. C., Hinkle, N.: J. Agric. Entomol. 2, 317 (1985).
303. Roush, R. T., Combs, R. L., Randolph, T. C., MacDonald, J., Hawkins, J. A.: J. Econ. Entomol. 79, 1178 (1986).
303a. McDonald, P. T., Schmidt, Ch. D., J. Econ. Entomol. 80, 433 (1987).
304. Shaw, T., Mayfield, R. J., Wimbush, J. M.: Proc. Int. Wool Text. Res. Conf. 1985, 225.
305. Farnham, A. W.: Pestic. Sci. 1985, 203.
306. Makino, S., Horikiri, H.: Kyushu Byogaichu Kenyugaiko 31, 175 (1985).
307. Keiding, J. P., Pederson, J. B.: Proc. Brit. Crop. Prot. Conf. 1986, 623.
308. Sawicki, R. M.: Intern. Entomol. Congr. Hamburg Sept. 1984.
309. Sawicki, R. M.: 190th ACS Meeting Chikago Sept. 1985.
310. Denholm, J., Farnham, A. W., O'Dell, K., Sawicki, R. M.: Bull. Entomol. Res. 73, 481 (1983).
310a. Farnham, A. W., Murray, W. A., Sawicki, R. M., Denholm, J., White, J. C.: Pestic. Sci. 1987, 209.
311. Matsunaga, T., Yoshida, K., Shinjo, G., Tsuda, G., Okano, Y., Yoshioka, M.: Pestic. Chem. Human Welfare Environ. Proc. IUPAC Congr. Kyoto 1982 III, 231.
312. Osborn, M., Hart, R. J.: Pestic. Sci. 1979, 407.
313. Miller, T. A., Kennedy, J. M., Collins, C.: Pestic. Biochem. Physiol. 12, 224 (1979).
314. Nicholson, R. A., Sawicki, R. M.: Pestic. Sci. 1982, 357.
314a. Brown, M. A.: J. Econ. Entomol. 80, 330 (1987).
315. Gammon, D. W.: Pestic. Biochem. Physiol. 13, 53 (1980).
316. Halliday, W. R., Georghiu, G. P.: J. Econ. Entomol. 78, 227 (1985).
317. Halliday, W. R., Georghiu, G. P.: J. Econ. Entomol. 78, 762 (1985).
318. Malcolm, C. A.: Genetica 60, 213 (1983).
319. Malcolm, C. A.: Genetica 60, 221 (1983).
320. Brealey, C. C., Crampton, P. C., Chadwick, P. R., Ricket, F. R.: Pestic. Sci. 1984, 121.
321. Chadwick, P. R., Slatter, R. S., Bowron, M. J.: Pestic. Sci. 1984, 112.
322. Scott, J. G., Matsumura, F.: Pestic. Biochem. Physiol. 19, 141 (1985).
323. Chialiang, C., Devonshire, A. L.: Pestic. Sci. 1982, 156.
324. De Vries, D. H., Georghiu, G. P.: Pestic. Biochem. Physiol. 15, 242 (1981).
325. Byford, A. R. L., Quisenberry, S. S., Spaks, T. C., Lockwood, J. A.: J. Econ. Entomol. 78, 768 (1985).
326. MacCaffery, A. R., Ahmed, M., Walker, C. H., Gladwell, R. F., Lee, K. S.: Proc. Brit. Crop. Prot. Conf. 1986, 583.
327. Collins, M. D.: Proc. Brit. Crop. Prot. Conf. 1986, 583.
328. Liu, M. Y., Tzeng, Y. J., Sun, C. N.: C.A. 94, 26116h (1986).
329. Chen, J. S., Lee, C. J., Yao, M. G., Sun, C. N.: J. Econ. Entomol. 78, 1198 (1987).
330. Chen, J. S., Sun, C. N.: J. Econ. Entomol. 79, 22 (1986).
331. Motoyama, N.: Pestic. Sci. 1984, 523.
332. McDonald, R. S., Soloman, K. R., Surgeoner, G. A., Harris, C. R.: Pestic. Sci. 1985, 10.
333. Golenda, C. F., Forgash, A. J.: J. Econ. Entomol. 77, 1105 (1985).
334. Pederson, L. E. K.: Experientia 42, 1057 (1986).
335. Scott, J. G., Georghiu, G. P.: Pestic. Sci. 1986, 195.

335a. Cilek, J. E., Knapp, F. W.: J. Agric. Entomol. *3*, 201 (1986).
336. Scott, J. G., Georghiu, G. P.: Biochem. Genet. *24*, 25 (1986).
337. Nicholson, R. A., Miller, T. A.: Pestic. Sci. *1985*, 561.
338. Ruzo, L. O., Casida, J. E., Gammon, D. W.: Pestic. Biochem. Physiol. *21*, 84 (1984).
338a. Krishnamurthy, V. V., Casida, J. E., Ruzo, L. O., J. Agric. Food Chem. *35*, 504 (1987).
339. Ghidiu, G. M., Carter, C. D., Silcox, C. A.: Proc. Brit. Crop. Prot. Conf. *1986*, 651.
340. Ishaaya, J., Casida, J. E.: J. Econ. Entomol. *10*, 681 (1981).
341. Riskallah, M. R.: Pestic. Biochem. Physiol. *19*, 184 (1983).
342. Chang, C. K., Whalon, M. E.: Pestic. Biochem. Physiol. *25*, 446 (1986).
343. Ho, S. H., Lee, B. H., See, D.: Toxicol. Lett. *19*, 127 (1983).
344. Chang, C. K., Jordan, D. W.: Pestic. Biochem. Physiol. *19*, 190 (1983).
344a. Scott, J. G., Mellon, R. B., Kirino, O., Gheorghiu, G. P.: Nippon Noyaku Gakaishi *11*, 475 (1986).
345. Schnitzerling, H. J., Nolan, J., Hughes, S.: Pestic. Sci. *1983*, 64.
345a. Nedelkina, S. V., Solomennikova, I. V., Volkotrub, E. N., Leonoa, I. N., Roslavtseva, S. A., Salganik, R. I.: Agrokhimiya *1987*, 103.
346. DuToit, G. O. G., Townsend, R. J., Armstrong, S. J.: N.Z.J. Exp. Agric. *6*, 175 (1978).
347. Devonshire, A. L., Moore, G. D.: Pestic. Biochem. Physiol. *18*, 235 (1982).
348. Farnham, A. W., Sawicki, R. M., White, J. C.: Proc. Brit. Crop. Prot. Conf. *1986*, 645.
349. Quisenberry, S. S., Lockwood, J. A., Byford, R. L., Wilsom, H. K., Sparks, T. C.: J. Econ. Entomol. *77*, 1095 (1984).
350. Lockwood, J. A., Byford, R. L., Story, R. N., Sparks, T. C., Quisenberry, S. S.: Environ. Entomol. *14*, 873 (1985).
351. Scirocchi, A., Militi, A.: Ann. Ist. Super Sanita *22*, 315 (1986).
352. Iftner, D. C., Hall, F. R., Sturm, M. M.: Pestic. Sci. *1986*, 242.
353. Elliott, M., Farnham, A. W., Janes, N. F., Johnson, D. M., Pulman, D. A., Sawicki, R. M.: Agric. Biol. Chem. *1986*, 1347.
354. Elliott, M., Janes, N. F., Farnham, A. W., Pulman, D. A., Needham, P. H.: Pestic. Sci. *1978*, 105.
355. Jaques, Y., Romey, G., Cavey, M. T., Kartalovski, B., Lazdunski, M.: Biochem. Biophys. Acta *600*, 882 (1980).
356. Narahashi, T.: Comp. Biochem. Physiol. *72*, 411 (1982).
357. Jackai, C. E. N., Sing, S. R.: Proc. Brit. Crop. Prot. Conf. *1986*, 761.
357a. Halliday, W. R., Feyereisen, R.: Pestic. Biochem. Physiol. *28*, 172 (1987).
358. Zoebelein, G.: (Bayer unpublished *1980*).
358a. Calliope (Beziers): technical information on Mamestrin® virus.
359. Terrier, L. C.: Ann. Rev. Entomol. *29*, 71 (1984).
359a. Dowd, P. F., Sparks, T. C.: Pestic. Biochem. Physiol. *27*, 123 (1987).
360. Nippon Kayaku: JA. 8516922 (1980).
360a. Dowd, P. F., Sparks, Th. C.: Pestic. Biochem. *27*, 237 (1987).
361. Rhone Poulenc: FR. 2485334 (1980).
362. Ciba Geigy: EP 96656 (1982).
363. Ishaaja, J., Ascher, K. R. S., Casida, J. E.: Abstr. Intern. Entomol. Conf. Hamburg 1982, R 16 – 1,3.
364. Koziol, F. J., Witkowski, J. F.: J. Econ. Entomol. *75*, 28 (1982).
365. Ishaaya, J., Ascher, K. R. S., Casida, J. E.: Crop. Prot. *2*, 335 (1983).
366. Gaughan, L. D., Ishaaya, J., Engel, J. L., Casida, J. E.: 190th ACS Meeting Chikago Sept. *1985*, Abstr., *119*.
367. Ube: JA 5896007 (1981).
368. Sankyo: JA 56120607 (1980).
369. Katsuda: JA 59210005 (1983).
370. ICI: AU 534419 (1979).
371. Sumitomo: JA 5064422 (1973).
372. Naumann, K.: in: Chemie der synthetischen Pyrethroid-Insektizide, Springer, Heidelberg 1981, p. 3.
373. Hatakoshi, M., Takayama, C., Matsuo, N., Nakayama, J., Kirino, O.: Nippon Noyaku Gakkaishi, *10*, 179 (1985).

374. Ghidiu, G. M., Silcox, E. A.: J. Agric. Entomol. *1*, 360 (1984).
375. Silcox, C. A., Ghidiu, G. M., Forgash, A. J.: J. Econ. Entomol. *78*, 1399 (1985).
376. Bayer: EP25179 (1979).
377. CSIRO: WO 843504 (1983).
378. Sumitomo: DAS 2348930 (1972).
379. Sumitomo: GB 2025771 (1978).
380. Sumitomo: NE 679854 (1967).
     Dainippon: JA 4839625 (1971); JA 4839626 (1971).
     Welcome Found: EP 5826 (1978).
381. Sumitomo: BE 805442 (1972), JA 5877529 (1973).
382. Roussel Uclaf: EP 17526 (1979).
383. Sumitomo: JA 806604 (1969).
384. Sumitomo: DOS 3447409 (1983).
385. Sumitomo: EP 148625 (1983).
386. Rajakulendran, S. V., Plapp, F. W.: J. Econ. Entomol. *75*, 1089 (1982).
387. Anda, M., Degheele, D.: Med. Fac. Landbouwwet. Rijksuniv. Gent *50*, 751 (1985).
388. Ishaaya, J., Yablonski, S., Ascher, K. S. R., Casida, J. E.: Phytoparasitica *12*, 99 (1984).
388a. Wellcome: EP 235979 (1986).
389. Morton, N., Byrne, J. E.: Proc. Brit. Crop. Prot. Conf. *1986*, 131.
390. Ahmed, J. M., Mostafa, A. M. A., Elewa, M. A.: J. Environ. Sci. Health B, *1985*, 689.
390a. Ochou, G., Hesler, L. S., Plapp, F. W.: Suppl. Southwest Entomol. *11*, 63 (1986).
390b. Wolfenbarger, D. A., Guerra, A. A.: ibid. *11*, 69 (1986).
390c. Gibson, R. W., Rice, A. D.: Ann. Appl. Biol. *109*, 465 (1986).
391. Sawicki, R. M.: in: Insecticides, ed. by Hudson, D. T., Roberts, T. R.: Wiley & Sons Ltd. 1985.
392. Pyrethrum Post *4*, 13 (1957).
393. Elliott, M.: J. Chem. Soc. *1964*, 888.
394. Godin, P.: J. Sci. Food Agric. *16*, 186 (1965).
395. Crombie, L.: Pestic. Sci. *1976*, 228.
396. Bullivant, M., Pattenden, G.: J. Chem. Soc. PT I *1976*, 249;
     Bullivant, M., Pattenden, G.: Pyrethrum Post *11*, 72 (1971).
397. Sumitomo: JA 5946243 (1982).
398. Bell, A., Kido, G.: J. Agr. Food Chem. *4*, 340 (1956).
399. Pyrethrum Post *2*, 11 (1950).
400. SCM Corp.: US 3839561 (1976).
401. Katsuda: JA 703827 (1967).
402. Sumitomo: JA 734780 (1970).
403. Sumitomo: JA 7446059 (1970).
404. Sumitomo: JA 4920325 (1972).
405. Sumitomo: DOS 2407403 (1973).
405a. Adams, M. E.: Vet. Res. Lab. PCT: WO 8603374 (1984).
405b. Mitsui Toatsu: JA 61152602 (1984).
406. Yoshitomi: JA 7218665 (1968).
407. Roussel Uclaf: US 4492687 (1980).
     Roussel Uclaf: US 4440756 (1981).
     Sakanke, S.: J. Pestic. Sci. *11*, 625 (1986).
408. Roussel Uclaf: DOS 3335360 (1982).
408a. Univ. of California: US 4622315 (1983).
408b. Chinoin: EP 147947 (1983).
409. Margulis, L., Rozen, H., Cohen, E.: Nature *315*, 658 (1985).
409a. Margulis, L., Cohen, E., Rozen, H.: Pestic. Sci. *1987*, 79.
410. Pieper, G. R., Rappoport, N. L.: Pyrethrum Post *15*, 104 (1984).
411. Dureja, P., Casida, J. E., Ruzo, L. O.: J. Agric. Food Chem. *1984*, 246.
414. Sumitomo: JA 5939807 (1982).
414a. Pest Controll *53*, 69 (1985).
415. Miskus, R. P., Andrews, T. L.: J. Agric. Food Chem. *20*, 313 (1972).
416. US 2967798 (1958).

417. Sumitomo: JA 4899327 (1972).
418. Sumitomo: JA 4926420 (1972).
419. Takeda: JA 4912027 (1972).
420. Maciver, D.: Pyrethrum Post *8*, 23 (1966).
421. Yamamoto, J., Katsuda, Y.: Pestic. Sci. *1980*, 134.
     Katsuda, Y., Hirobe, H., Minamite, Y.: Pestic. Chem. Human Welfare Environ. Proc.
     IUPAC Congr. Kyoto 1982 II, 223.
422. Teijin: JA 55149202 (1979); JA 55149203 (1979).
422a. Adams Vet. Res. Lab.: US 4668666 (1984).
423. Dainippon: JA 53107419 (1977).
424. Penwalt: EP 172934 (1984).
425. Penwalt: US 4497793 (1980).
426. Otsuka: JA 6130505 (1985).
427. Sumitomo: DOS 2924878 (1978); DOS 2924879 (1978).
428. Sumitomo: JA 8146802 (1979).
429. Shell: EP 29626 (1979).
430. Ciba Geigy: EP 74335 (1981).
431. Bayer: DOS 2707232 (1977).
432. ACC: Braz. Ped.: BR 825599 (1981).
433. ICI: BE 861368 (1976).
434. Penwalt: WO 80994 (1984).
     Adams: WO 863374 (1984).
435. Shell: NE 178556 (1978).
436. Fearing Manuf. Comp.: EP 143297 (1983).
437. Bayer: DOS 3333657 (1983).
438. Lion Dentifrice: JA 5538326 (1978).
439. Sumitomo: JA 60169403 (1984).
440. Roussel Uclaf: GB 2065475 (1979).
441. Welcome: GB 2088212 (1980).
442. Janssen: EP 61208 (1981).
443. Bayer: DOS 3208334 (1982).
444. Wellcome: AU 8433208 (1984).
445. Janssen: GB 2148119 (1984).
446. Liebisch, A.: Veterinär-Medizinische Nachrichten *1986*, 17.
447. Stendel, W.: Veterinär-Medizinische Nachrichten *1986*, 28.
448. Hamel, H. D., Amelsfoort, A. V.: Veterinär-Medizinische Nachrichten *1986*, 34.
449. Dorn, H., Romano, A., Pulga, M., Rodriguez, O.: Veterinär-Medizinische Nachrichten
     1986, 40.
449a. Jenkinson, D. M., Hutchinson, G.: Research in Veterinary Sci. *1986*, 237.
450. ACC: DOS 2601743 (1975).
450a. Quercy: US 4624070 (1983).
451. Kuruyama: JA 7122239 (1964).
452. Solimpex: FR 2446858 (1979).
452a. Poudres at Explosives EP 217696 (1985).
453. Teijin: JA 55149202 (1979).
     Okada: JA 6233102 (1985).
454. Matsumoto: JA 54151127 (1978).
455. Dainippon: JA 6075403 (1983).
456. Katsuda: JA 54123371 (1978).
457. Globol: DOS 3436310.
458. Airwick: GB 2070431 (1980).
459. Budapest Muszaki: HU T 28623 (1981).
460. Briggs, G. G., Elliott, M., Janes, N. F.: Pestic. Chem. Human Welfare Environ. Proc.
     IUPAC Congr. Kyoto 1982 I, 157.
461. Proc. Brit. Crop. Prot. Conf. *1986*, 98; 205; 1110.
462. Elliott, M., Janes, N. F. et al.: Pestic. Sci. *1976*, 236.
463. Hull, L. A., van Starner, R.: J. Econ. Entomol. *76*, 122 (1983).

464. Boslanian, N. J., Belanger, A.: Int. Entomol. Conf. Hamburg 1982, Abstr. R17–1,7.
465. Croft, B. A., Wagner, S. W.: J. Econ. Entomol. 74, 703 (1981).
466. Wong, S. W., Chapman, R. B.: Aust. J. Agric. Res. 30, 497 (1979).
467. Abu-Awad, B. A., Banhawy, E. M.: Exp. Appl. Acarol. 1, 185 (1985).
468. Cranham, J. E., Easterbrook, M. A., Kapetanakis, E.: Pestic. Sci. 1985, 214.
469. Everts, J. W., Kortenhoff, B. A., Hochland, H., Vlug, H. J., Joque, J., Koeman, J. H.: Arch. Environ. Contam. Toxicol. 14, 641 (1985).
470. Basedow, T., Rzehak, H., Voß, K.: Pestic. Sci. 1985, 325.
471. Cole, J. F. H., Everett, C. J., Wilkinson, E., Brown, R. A.: Pestic. Sci. 1985, 212.
472. Basedow, T.: Proc. Brit. Crop. Prot. Conf. 1986, 181.
473. Delornine, R., Berthier, A., Auge, D.: Pestic. Sci. 1985, 332.
474. Powell, J. E., King, E. G., Jany, C. S.: J. Econ. Entomol. 79, 1343 (1986).
475. Tewasi, G. C., Moorthy, P. N. K.: Indian. J. Agric. Sci. 55, 40 (1985).
476. Focks, D. A.: Mosq. News 44, 534 (1984).
477. Plapp, F. W., Vinson, S. B.: J. Econ. Entomol. 6, 381 (1977).
478. Rajakulendran, S. V., Plapp, V. W.: J. Econ. Entomol. 75, 769 (1982).
479. Greidanus: EP 169621 (1984).
480. Elliott, M., Janes, N. F., Stevenson, H., Walters, H. H.: Pestic. Sci. 1983, 425.
481. Inglesfield, C.: Pestic. Sci. 1983, 211.
482. Eremina, O. Y., Roslavtseva, S. A., Sobchak, M. N., Sabirova, D.: Agrokhimiya 11, 108 (1985).
483. Arzone, A., Vidano, C.: Apic. Mod. 76, 203 (1985).
484. Garnier, A. P., Roa, C., Herve, J. J.: Def. Veg. 231, 32 (1985).
485. Elliott, M., Janes, N. F., Farnham, A. W., Potter, C.: Annu. Rev. Entomol. 23, 443 (1978).
486. Adkins, A. L., Kellum, D., Adkins, K. W.: Am. Bee J. 118, 542 (1978).
487. Pike, K. S., Mayer, D. F., Glazer, M., Kions, C.: Environ. Entomol. 11, 951 (1982).
488. Smart, L. E., Stevenson, J. H.: Bee World 63, 150 (1982).
489. Moffet, J. O., Stoner, A., Ahring. R. M.: South-West Entomol. 7, 111 (1982).
490. Shires, S. W., Murray, A., Debray, P., LeBlanc, J.: Pestic. Sci. 1984, 491.
491. Shires, S. W., LeBlanc, J., Murray, A., Forbes, S., Debray, D.: J. Apic. Res. 23, 217 (1984).
491a. Arzone, A., Vidano, C.: Apic. Mod. 76, 203 (1985).
492. Fries, Z.: Vaextskydds Rapp. Jordbruk 32, 169 (1985).
493. Shires, S. W.: Pestic. Sci. 1985, 215.
494. LeBlanc, J.: Pestic. Sci. 1985, 206.
495. Gehring, L.: Pestic. Sci. 1985, 206.
496. Tasai, J. N., Debray, D.: Pestic. Sci. 1985, 209.
497. Delabie, J., Bos, C., Fonta, C., Masson, C.: Pestic. Sci. 1980, 409.
497a. Bayer: EP 224697 (1985).
498. Nijhuis: DOS 3503255 (1985).
499. Danka, R. G., Rinderer, T. E., Hellmich, R. L., Collins, A. M.: J. Econ. Entomol. 79, 18 (1986).
500. Yu, S. J., Robinson, F. A., Nation, J. L.: Pestic. Biochem. Physiol. 22, 360 (1984).
501. Bounias, M., Popeskovich, D. S.: Pestic. Biochem. Physiol. 24, 149 (1985).
502. Singh, J., Sidhu, A. S.: J. Entomol. Res. 9, 7 (1985).
503. Wilkinson, W., Gough, H. J., Collins, J. G.: Proc. Brit. Crop. Prot. Conf. 1986, 1085.
504. The Pyrethroid Insecticides Ed. Leahey, J. P.: by Taylor and Francis, London/Philadelphia 1985.
505. Jeffery-O'Donnel, N. L.: Bull. Environ. Contam. Toxicol. 23, 250 (1979.
506. Kahn, N. J.: Pestic. Chem. Human Welfare Environ. Proc. IUPAC Congr. Kyoto 1982 III, 437.
     Kreutzweiser, D. P., Kingsbury, P. D.: Pestic. Sci. 1987, 49.
506a. Natschin, Y. V., Lavrova, E. A., Khripak, A. V.: Vopr. Ikhtol. 26, 998 (1986).
507. Ishimitsu, K., Kasahara, J., Yamada, T., Soma, S., Kamimura, H.: Pestic. Chem. Human Welfare Environ. Abstr. IUPAC Congr. Kyoto 1982 Ia-4.
507a. Mikami, N., Sakata, S., Yamada, H., Miyamoto, J.: J. Pestic. Sci. 9, 697 (1984).
508. Nippon Soda: DOS 3033358 (1979).

509. Ide, J., Nakada, Y., Endo, R., Muramatsu, S., Konishi, K., Mizuno, T., Ohno, S., Yamazaki, Y., Endo, H.: Agric. Biol. Chem. *47*, 927 (1983).
510. Kuraray: DOS 3145448 (1980).
511. Hotellier, M. L., Vincent, P.: Proc. Brit. Crop. Prot. Conf. *1986*, 1109.
512. Crossland, N. O.: Aquat. Toxicol. *2*, 205 (1984).
513. Smies, M., Evers, R. H. J., Peijnenburg, F. H. M., Koeman, J. K.: Ecotoxicol. Environ. Saf. *4*, 114 (1980).
514. Shires, S. W.: Bull. Environ. Contam. Toxicol. *34*, 134 (1985).
515. Muir, D. C. G., Rawn, G. P., Grift, N. P.: J. Agric. Food Chem. *33*, 603 (1984).
516. Allied Colloids: EP 156623 (1984).
517. Stephenson, R. R.: Pestic. Sci. *1985*, 199.
518. Glickmann, A. H., Lech, J. J.: Toxicol. Appl. Pharmacol. *1981*, 186.
519. Bocquet, J. C., Hotellier, M. L.: Pestic. Sci. *1985*, 198.
520. Kingsbury, P. D., Kreuzweiser, D. P.: Pestic. Sci. *1985*, 202.
521. Komalah, Z.: Folia. Biol. (Krakow) *33*, 9 (1985).
522. Cole, L. M., Casida, J. E.: Pestic. Biochem. Physiol. *20*, 217 (1983).
523. Fichera, L. E., Salibian, A., Rodolfo, R. R.: Comp. Physiol. Ecol. *11*, 1 (1986).
524. ICI: GB 1537499 (1975).
525. BASF: DOS 2704962 (1977).
526. Ciba Geigy: GB 1551852 (1977).
527. Roussel Uclaf: FR 2461458 (1979).
528. Crammer, B., Goldschmidt, Z., Ikan, R., Spiegelstein, H.: J. Agric. Food Chem. *1985*, 1148.
529. JL 70575 (1983).
530. Stauffer: DOS 2114460 (1970).
531. Singh, O. P., Varma, P. K., Singh, B. R.: Pesticides *21*, 49 (1985).
532. Wood, B. W., Payne, J. A.: Host. Sci. *21*, 112 (1986).
533. Sumitomo: DOS 2737297 (1976).
534. Marder, E. O., Udey, E. C.: Phytopathology *27*, 112 (1937).
535. DOW: US 4423222 (1982).
536. Wild, A., Oberweiß, A. L., Röhle, W.: Z. Pflanzenphysiol. *82*, 161 (1977).
537. Moreland, D. E., Noritzky, W. P.: Adv. Photosynth. Res. Proc. Int. Cong. Photosynth. 1984, 81.
538. Chauhan, L. K. S., Dikshit, T. S. S., Sundararaman, V.: Mutat. Res. *141*, (1986) 25.
539. Roussel-Uclaf: DOS 3207009 (1981).
540. Roussel-Uclaf: EP 38271 (1980).
541. Roussel-Uclaf: EP 107570 (1982).
542. Tu, C. M.: Environ. Sci. Health, B, *17*, 43 (1982).
543. Draughon, A., Ayres, J. C.: J. Agric. Food Chem. *28*, 1115 (1980).
544. Stratten, G. W., Korke, C. T.: Environ. Polut. Ser. A *29*, 71 (1982).
545. Tu, C. M.: Microb. Ecol. *5*, 321 (1980).
545a. Chapman, R. A., Harris, C. R.: J. of Chromatogr. *174*, 369 (1979).
545b. Roberts, T. R., Standen, M. E.: Pestic. Sci. *1977*, 305.
545c. Chapmann, R. A., Tu, C. M., Harris, C. R., Cole, C.: Bull. Environ. Contam. Toxicol. *26*, 513 (1981).
546. Roussel-Uclaf: FR 2579867 (1985).
546a. Carle, P. G., Coz, J., Elissa, N., Gasquet, M.: C.R. Acad. Sci. III, *303*, 565 (1986).
547. Pestic. Residues in Food FAO Evaluations 1979, 1980, 1981.
547a. Bayer: DOS 3123610 (1981).
548. Kavlok, R., Chernoeff, N., Baron, A., Lindner, R., Rogers, R.: J. Environ. Pathol. Toxicol. *2*, 751 (1979).
549. Chanh, P. H., Navarro-Delmasure, C., Chanh, A. P., Martinez, C.: IRCS Med. Sci. Libr. Compend. *9*, 565 (1981).
550. Cyfluthrin: Technical Information Bayer AG.
551. James, J. A.: Wellcome Res. Lab. Berghamsted.
551a. Fastac: Technical Prospect, Shell 1986.
553. Gray, A., Connors, T. A.: Soc. of Chem. Ind. Pestics. Group-Meeting Med. Res. Lab. Counc. Lab. Carshalton 29. 9. 78.

554. Cremer, J. E., Ray, D. E.: abstr. Neurotox. Symposium Bath *1985*, 45.
555. Gray, A. J., Connors, T. A., Hoellinger, Harm, N. H.: Pestic. Biochem. Physiol. *13*, 281 (1980).
556. Gray, A. J., Connors, T. A.: Pestic. Sci. *1980*, 361.
557. Gray, A. J., Connors, T. A., Richard, J.: Neurotrans., Their Recept. Workshop *1980*, 565.
558. Forshaw, P. J., Ray, D. E.: Pestic. Biochem. Physiol. *25*, 143 (1986).
559. Verschoyle, R. D., Aldridge, W. N.: Arch Toxicol. *45*, 325 (1980).
559a. Barnes, J. M., Verschoyle, R. D.: 1972–1974.
560. Elliott, M.: Paper given at the CILDA-Meeting, April 1976.
561. Lawrence, L. J., Casida, J. E.: Pestic. Biochem. Physiol. *18*, 9 (1982).
562. Verschoyle, R. D., Aldridge, W. N.: Arch. Toxicol. *45*, 325 (1980).
563. Cremer, J. E., Ray, D. E.: Abstr. Neurotox' 85, Symposium Barth, April 1985, p. 45.
564. Brodie, M. E.: Neurobehav. Toxicol. Terratol. *7*, 51 (1985).
565. Gray, A., Connors, D. A., Rickard, J.: Neurotrans., Their Recept. Workshop *1980*, 565.
566. Staatz-Benson, C. G., Hosko, M. J.: Pestic. Biochem. Physiol. *25*, 19 (1985).
567. Honerjager, P.: Rev. Physiol. Biochem. Pharmacol. *92*, 1 (1982).
568. Gray, A. J., Rickard, J.: Pestic. Biochem. Physiol. *18*, 205 (1982).
569. Rickard, J., Brodie, M. E.: Pestic. Biochem. Physiol. *23*, 143 (1985).
570. Lock, A. E., Berry, P. N.: Dev. Toxicol. Environ. Sci. *8*, 623 (1980).
570a. Morean, R., Carle, P. R., Gourdoux, L., Couilland, F., Khay, A. B., Girardie, A.: Pestic. Biochem. Physiol. *27*, 101 (1987).
571. Ray, D. E.: Exp. Brain Res. *45*, 269 (1985).
572. Cremer, J. E., Seville, M. P.: Neurotoxicology *6*, 1 (1985).
573. Cremer, J. E., Cunningham, V. J., Ray, D. E.: Brain Res. *194*, 278 (1980).
574. Hutson, D. H., Gaughan, L. C., Casida, J. E.: Pestic. Sci. *1981*, 385.
574a. Hesler, L. S., Plapp, F. W.: Suppl. South West Entomol. *11*, 75 (1986).
574b. Sumitomo: GB 2176107 (1985).
575. Brodie, M. E., Aldridge, W. N.: Neurobehav. Toxicol. Teratol. *4*, 109 (1982).
576. Soderlund, D. M., Casida, J. E.: Pestic. Biochem. Physiol. *7*, 391 (1977).
576b. Edwards, R., Milburn, P., Hutson, D. H.: Pestic. Sci. 1987, 1.
576c. Ueda, K., Ganghan, C. C., Casida, J. E.: Pestic. Biochem. Physiol. *1975*, 280.
577. Brodie, M. E.: Neurotoxicology *4*, 1 (1983).
578. Rao, C. R., Motoyama, N., Dautermann, W. C.: Pestic. Biochem. Physiol. *23*, 66 (1985).
579. Dyball, R. E. J.: Pestic. Biochem. Physiol. *17*, 42 (1982).
580. Ahmed, F. A., Mohamed, A. Z., Ismail, S. A., Maran, A.: Egypt. J. Food Sci. *12*, 85 (1984).
581. Gray, A. J., Connors, T. A., Hoellinger, H., Nguyen, H. N.: Pestic. Biochem. Physiol. *13*, 281 (1980).
582. Riviere, J. L., Bach, J., Grolleau, G.: Bull. Environ. Contam. Toxicol. *31*, 479 (1983).
583. Carlson, G. P., Schoeningh, G. P.: Toxicol. Appl. Pharmacol. *52*, 507 (1980).
584. Habazin-Novak, V., Plestina, R.: Period. Biol. *86*, 315 (1984).
585. LeQuesne, P. M.: Neurol. Neurosurg. Psychiatrie *85*, 1005 (1982).
586. Pham, H. C., Navarro Delmasure, C., Pham, A. H. C., Lean, S. C., Ziade, F., Samaha, F.: IRCS Med. Sci. Libr. Compend. *9*, 503 (1981).
587. Strohmeyer, A., Karmos-Varszegi, M., Szlobodnyik, J., Zilahy, G., Dula, G., Torok, G., Desi, J.: Egeszsegtudomany *28*, 62 (1984).
588. Bloom, A. S., Staatz, C. G., Diehringer, T.: Neurobehav. Toxicol. Teratol. *5*, 321 (1983).
589. Litchfield, M. H.: Pestic. Chem., Human Welfare Environ. Proc. IUPAC Congr. Kyoto 1982 III, 206.
590. Nagy, K., Bedo, M., Bajzath, J., Szepvolgyi, J., Antal, M.: C.A. 100, 169617a (1983).
591. Rose, G. P., Dewar, A. J.: Arch. Toxicol. *53*, 297 (1983).
592. Aldridge, W. N.: Pestic. Chem. Human Welfare Environ., Proc. IUPAC Congr. Kyoto 1982 III, 485.
593. Dyck, P. J., Chimono, M., Schoening, G. P., Lais, A. C., Oviatt, K. F., Sparks, M. F.: J. Environ. Pathol. Toxicol. Onkol. *5*, 109 (1984).
594. Rhodes, C.: Pestic. Sci. *1984*, 471.
594a. Casida, J. E., Ruzo, L. O.: in Natural Products for Innovative Pest Management p. 111 Ed. by Whitehead a. Bowers, Pergamon Press 1983.

594b. Casida, J. E., Coretta, A., Gaughan, C., Ruzo, L. O.: Advance in Pestic. Sci. Proc. IUPAC Congr. Zürich 1978 II, 183.
594c. Casida, J. E., Unai, T.: J. Agric. Food Chem. 25, 979 (1977).
594d. Casida, J. E., Soderlund, D. M.: ACS Symp. Ser. 42, 173 (1977).
594e. Shono, T., Casida, J. E.: J. Pestic. Sci. 3, 165 (1978).
594f. Leahey, J. P.: in The Pyrethroid Insecticides. Ed. Leahey by Taylor & Francis 1985, 263–342.
594g. Soderlund, D. M., Casida, J. E.: Pestic. Biochem. Physiol. 7, 391 (1977).
596. Soderlund, D. M., Ghiasuddin, S. M.: Pestic. Biochem. Physiol. 24, 200 (1985).
597. Holan, G., Frelin, C., Lazdunski: Experientia 41, 520 (1985).
598. Lawrence, L. J., Casida, J. E.: Pestic. Biochem. Physiol. 18, 9 (1982).
599. Soderlund, D. M.: 190th ACS Meeting, Chikago Sept. 1985.
600. Nicholson, R., Wilson, R., Potter, C., Black, M.: Pestic. Chem. Human Welfare Environ. Proc. IUPAC Congr. Kyoto 1982 III, 75.
601. Rao, K. S. P., Chetty, S. C., Desaiah, D.: J. Toxicol. Environ. Health 14, 257 (1984).
602. Berlin, J. R., Akera, T., Brodie, T. M., Matsumura, F.: Eur. J. Pharmacol. 89, 313 (1984).
603. Searle: US 3098857 (1961).
604. Forshaw, P. J., Bradbury, J. E.: Eur. J. Pharmacol. 91, 207 (1983).
605. McCorkle, F., Taylor, A., Martin, D., Glick, B.: Poultry Science 59, 1568 (1980).
606. Desi, J., Varga, L., Dobronyi, J.: Arch. Toxicol. Suppl. 8, 305 (1985).
607. Stelzer, J. K., Gordon, A. M.: Res. Commun. Chem. Pathol. Pharmacol. 46, 137 (1984).
608. Desi, J., Dobronyi, J., Varga, G.: Egeszsegtudomany 30, 86 (1986).
608a. Desi, J., Dobronyi, J., Varga, L.: Ecotoxicology and Environ. Safety 12, 220 (1986).
609. David, D.: Poultry Science 60, 1149 (1981).
610. New Scientists 4/1986.
611. Flannigan, S. A., Tucker, S. B., Key, M. M., Ross, C. E., Fairchild, E. J., Grimes, A. B., Harrist, R. B.: Arch. Toxicol. 56, 288 (1985).
612. Cagen, S. Z., Malley, L. A., Parker, C. M., Gardiner, T. H., Gelder, G. A. V., Jud, V. A.: Toxicol. Appl. Pharmacol. 76, 270 (1984).
613. Bainova, A., Kaloyanova, F.: Khig. Zdraveopaz. 28, 19 (1985), C.A. 103, 208255g.
613a. McKillop, C. M., Brock, J. A., Oliver, G. J. A., Rodes, C.: Toxicol. Lett. 36, 1 (1987).
614. Flannigan, S. A., Tucker, S. B.: Contact. Dermatitis 13, 140 (1985).
615. Tucker, S. B.: Arch. Toxicol. 54, 195 (1983).
616. Kolmodin-Hedman, B., Svensson, A., Aakerbloom, M.: Arch. Toxicol. 50, 27 (1982).
617. He, F., Han, K., Yao, P., Wu, T., Wang, S., Li, L., Yang, H., Sun, B., Zhang, X.: Zhoughuo Laodzong Weisheng, Zhiyebing Zazhi 4, 72 (1986) C.A. 105, 158134t.
618. Eadsforth, C. V., Baldwin, M. U.: Xenobiotica 1983, 67.
619. Croucher, A., Logan, C. J.: 184th ACS Meeting Kansas City 8/82 Pestic. Div. No. 39.
620. FMC-Study: Pesticides and Toxic. Chem. News 9, 28 6/1981.
620a. Agrichemical Age Jan. 1987, 12a.
621. Ruzo, L. O., Casida, J. E.: Environ. Health Perspectives 21, 285 (1977).
621a. Gray, A. J., Soderlund, D. M. in: Insecticides, Ed. by Hudson a. Roberts Wiley and Sons 1985.
622. Miyamoto, J.: Environ. Health Perspectives 14, 15 (1976).
623. Nishi, K., Nunoshiba, T., Nishika, H.: Sci. Eng. Rev. Doshisha Univ. 26, 93 (1985); C.A. 104 63367M.
624. Ding, D., Yu, Y., Zhang, J., Cai, Zh., Chen, X.: Zejian Yike Daxue Yuebao 14, 1 (1985); C.A. 103 155575k.
625. Mayberry, R. N., Savage, J.: Abstr. Ann. Meet. Amer. Soc. Microbiol. 78, 126 (1978).
626. Kimmel, E. C.: J. Agric. Food Chem. 30, 623 (1980).
627. Klopman, G., Contreiras, R., Rosenkranz, H. S., Waters, M. D.: Mutation Res. 147, 343 (1985).
628. Casida, J. E., Ruzo, L. O.: J. Agric. Food Chem. 1982, 623.
629. Simmon, V. F., Riccio, E. S., Robinson, D. E., Mitchell, A. D.: Gov. Rep. Announce Index (US) 1985, 85 (16) 535786 C.A. 104, 201920Q.
630. Gaughan, L. C., Ishaaya, I., Engel, J. L., Casida, J. E.: Abstr. 190th ACS Meeting, Chikago Sept. 1985; 119.

631. Devaud, L. L., Szot, B., Murray, Th. F.: Eur. J. Pharmacol. *121*, 269 (1986).
632. Pham, C. H., Navarro-Delmasure, C., Pham, A. C. H., Cheav, S. L., Ziade, F., Samaha, F.: Arzneimittelforsch. *34*, 175 (1984).
633. Nishimura, M., Obana, N., Yagasaki, O., Yanagiya, J.: Toxicol. Sci. *9*, 131 (1984).
634. Staatz, G. C., Bloom, A. S., Lech, J. J.: Pestic. Biochem. Physiol. *17*, 287 (1982).
635. Gammon, D. W., Lowell, J., Casida, J. E.: Toxicol. Appl. Pharmacol. *66*, 290 (1982).
636. Machemer, L., Eben, A., Thyssen, J.: (Bayer) Paper delivered at the Toxicology Conf. Alexandria Nov. 1983.
637. Cremer, J. E., Cunningham, V. J., Ray, D. E., Sarna, S. G.: Brain Res. *194*, 378 (1984),
638. Pham, H. C., Navarro-Delmasure, C., Pham, A. C. H., Clavel, P., Gayrel, P.: IRCS Med. Sci. Libr. Compend. *8*, 388 (1980).
639. Bradburry, J. E., Forshaw, P. J., Gray, A. J.: Neuropharmacol. *22*, 907 (1983).
640. Tilson, H. A., Hong, J. S., Mactutus, G. F.: J. Pharmacol. Exp. Ther. *233*, 285 (1985).
641. Horimori, T., Nakanishi, T., Kawaguchi, S., Sako, H., Suzuki, T., Miyamoto, J.: Nippon Noyaku Gakkaishi *11*, 9 (1986).
642. Bradburry, J. E., Gray, A. J., Foreshaw, P.: Appl. Toxicol. *60*, 382 (1981).
643. Leclerque, M., Cotonat, J., Foulhoux, P.: J. Toxicol. Clin. Exp. *6*, 85 (1986).
644. Thiebault, J. J., Bort, R., Foulhoux, P.: Coll. Med. Leg. Toxicol. Med. *47*, 131 (1985).
645. Terriere, L. C.: Ann. Rev. Entomol. *29*, 71 (1984).
646. Sumitomo: JA 9093018 (1982).
646a. Shell: US 4451482 (1984).
647. = 646a.
648. Golenda, C. T., Forgash, A. J.: Entomol. Exp. Appl. *40*, 169 (1986).
649. Gammon, D. W.: Pestic. Sci. *1978*, 79.
650. Gammon, D. W.: Pestic. Biochem. Physiol. *7*, 1 (1977).
651. Riordan, E. K.: Entomol. Exp. Appl. *1986*, 193.
652. Adams, M. E., Miller, T. A.: Proc. Neurotox' 79, 430 Symposion York Sept. 1979.
653. Nakagawa, S., Okajima, N., Kitahaba, T., Nishimura, A., Fujita, T., Nakayima, U.: Pestic. Biochem. Physiol. *17*, 243; 259 (1982).
654. Greenwood, R., Marshman, S. A., Ford, M. G.: Abstr. Neurotox' 85, 122 Symposion Barth April 1985.
655. Loewenskin, O.: Nature *150*, 760 (1942).
656. Berteau, P. E., Casida, J. E., Narahashi, T.: Sci. *161*, 1151 (1968).
657. Narahashi, T.: Bull. Wld. Health Org. *44*, 7 (1971).
658. Leake, L.: Pestic. Sci. *1977*, 713
659. Burt, P. E., Goodchild, R. E.: Entomologia Exp. Appl. *14*, 179 (1971).
660. Burt, P. E., Goodchild, R. E.: Entomologia Exp. Appl. *14*, 255 (1971).
661. Miller, T., Adams, M. E.: Synthetic. Pyrethroids ACS Symposium Ser. *42*, 98 (1977).
662. Narahashi, T., Nishimura, K.: Synthetic. Pyrethroids ACS-Symposium Ser. *42*, 85 (1977).
663. Takeno, K., Nishimura, K., Parmentier, J. L., Narahashi, T.: Pestic. Biochem. Physiol. *7*, 486 (1978).
664. Nishimura, K., Narahashi, T.: Pestic. Biochem. Physiol. *8*, 53 (1978).
665. Burt, P. E., Goodshild, R. E.: Pestic. Sci. *1977*, 681.
666. Brigga, G. G., Elliott, M., Farnham, A. W., Janes, N. F.: Pestic. Sci. *1974*, 643.
667. Martell, J. J.: 190th ACS Meeting, Chikago Sept. 1985.
678. Bercken, van den, J., Vijverberg, H.: Proc. Neurotox' 79, 391, Symposion York Sept. 1979.
669. Narahashi, T., in: Pyrethroid Insecticides; Table Ronde Roussel-Uclaf *37*, 15 (1980).
670. Gammon, W. D., Brown, M. A., Casida, J. E.: Pestic. Biochem. Physiol. *15*, 181 (1981).
671. Gammon, W. D., Ruzo, L. O., Casida, J. E.: Neurotoxicology *4*, 165 (1983).
672. Lauffer, J., Roche, M., Pelhate, M., Elliott, M., Janes, N. F., Satelle, D. B.: J. Insect. Physiol. *30*, 341 (1984).
673. Roche, M., Benoit, M., Herve, J. J. R., Carle, P. R.: Proc. Brit. Crop. Prot. Conf. *1984*, 937.
674. Narahashi, T.: Pestic. Chem. Human Welfare Environ. Proc. IUPAC Congr. Kyoto 1982 II, 109.
675. Lawrence, L. J., Casida, J. E.: Pestic. Biochem. Physiol. *18*, 9 (1982).
676. Salgado, V. L., Irving, S. N., Miller, T. A.: Pestic. Biochem. Physiol. *20*, 169 (1983).
677. Irving, S. N.: Proc. Brit. Crop. Prot. Conf. *1984*, 859.

678. MacDonald, E., Punja, N.: Pestic. Sci. *1986*, 459.
679. Nishimura, K., Ueno, A., Nakagava, S., Fujita, T., Nakayima, M.: Pestic. Biochem. Physiol. *17*, 271 (1982).
680. Omatsu, N., Nishimura, K., Fujita, T.: Pestic. Biochem. Physiol. *25*, 288 (1986).
681. Scott, J. G., Matsumura, F.: Pestic. Biochem. Physiol. *19*, 141 (1983).
682. Leake, L. D., Buckley, D. S., Ford, M. E., Salt, D. W.: Neurotoxicology *6*, 99 (1985).
683. Hendy, C. H., Djamgoz, M. B.: Pestic. Sci. *1985*, 520.
684. Greene, R., Hawes, A. N., Leake, L. D.: Abstr. Neurotox' 85, 120 Symposion Bath April 1985.
685. Ford, M. G., Greenwood, R., Salt, D. W., Moss, M., Chalmers, A. E.: Abstr. Neurotox '85, 116, Symposion Bath April 1985.
686. Pichon, Y., Guillet, J. C., Heilig, U., Pelhate, M.: Pestic. Sci. *1985*, 627.
687. Buckley, D. S., Leake, L. D., Ford, M. G.: Pestic. Sci. 1985, 545.
688. Heilig, U., Pichon, Y.: Abstr. Neurotox' *85*, 76 Symposion Bath April 1985.
689. Roche, M., Guillet, J. C.: Pestic. Sci. *1985*, 511.
690. Murayama, K., Abbott, N. J., Narahashi, T., Shapiro, B. J.: Comp. Gen. Pharmacol. *3*, 391 (1972).
691. Narahashi, T.: Pestic. Sci. *1976*, 267.
692. Narahashi, T., Lund, A. E.: Proc. Neurotox' *79*, 497 Symposion York Sept. 1989.
693. Bashford, C. L., Pasternak, C. A.: TIBS-Letters *1986* 113.
694. Leake, L. D.: Paper given at Neurotox' 85 Symposion Bath, April 1985.
695. Ritchie, R. M.: Abstr. Neurotox' 85, 130, Symposion Bath April 1985.
696. Narahashi, T.: Paper given at the Neurotox' *85*, Symposion Bath, April 1985.
697. Neumke, B.: Naturwissenschaften *72*, 17 (1985).
698. Gammon, D. W., Sander, G.: Neurotoxicology *6*, 63 (1985).
699. McDonald, E.: Paper given at the 190th ACS Meeting Chikago Sept. 1985, abstr. p. 187.
700. Ruzo, L. O., Casida, J. E., Gammon, D. W.: Pestic. Biochem. Physiol. *21*, 84 (1984).
701. Noda, M., Ykeda, T., Kayano, T., Suzuki, H., Takeshima, H., Kurasaki, M., Takahashi, H., Numa, S.: Nature *320*, 188 (1986).
701a. Weiß, D. E.: Aust. J. Biol. Sci. *22*, 1355 (1969).
702. Rashatwar, S., Matsumura, F.: Comp. Pharmacol. Toxicol. *81*, 97 (1985).
703. Lauffer, J., Pelhate, M., Sattelle, D. B.: Pestic. Sci. *1985*, 651.
704. Narahashi, T.: Abstr. Neurotox' *85*. 225, Symposion Bath April 1985.
705. Lauffer, J., Roche, M., Pelhate, M., Elliott, M., Janes, N. J., Satelle, D. B.: Insect. Physiol. *30*, 341 (1984).
706. Osborne, M. P., Smallcombe, A.: Pestic. Chem. Human Welfare Environ. Proc. IUPAC Congr. Kyoto 1982 I, 103.
707. Burt, P. E.: Proc. Neurotox' *79*, 407, Symposion York Sept. 1979.
708. Osborne, M. P.: Proc. Neutrotox' *79*, 37, Symposion York Sept. 1979.
710. Jaques, Y., Romey, G., Cavey, M. T., Lazdunski, M.: Biochem. Biophys. Acta *600, 882* (1980).
711. Catterall, W. A., Risk, M.: Molecular Pharmacol. *19*, 345 (1981).
712. Narahashi, T.: 190th ACS Meeting, Chikago Sept. 1985.
713. Pichon, Y., Pelhate, M.: 190th ACS Meeting, Chikago Sept. 1985.
714. Elliott, M., Farnham, A. W., Janes, N. F., Johnson, D. M., Pulman, D. A., Sawicki, R. F.: Agric. Biol. Chem. *1986*, 1347.
715. Nicholson, R. A., Blade, R. J., Botham, R. R.: Abstr. Neurotox' *85*, 133, Symposion Bath, April 1985.
716. Farnham, A. W., Sawicki, R. M., White, J. C.: Proc. Brit. Crop. Prot. Conf. *1986*, 645.
716a. Narahashi, T., Moore, H. W., Shapiro, B. J.: Sci. *163*, 680 (1969); Narahashi, T., Seyama, I.: J. Physiol. *242*, 417 (1974).
716b. Narahashi, T., Shapiro, B. J., Deguchi, T., Scriba, M., Wang, C. M.: Am. J. Physiol. *222*, 850 (1972).
716c. Murayama, K., Narahashi, N. J., Shapiro, B. J.: Comp. Gen. Pharmac. *3*, 391 (1972).
717. Bergmann, C., Dubois, J. M., Rojas, E., Rethmayer, W.: Biochimica and Biophysica Acta *455*, 173 (1976).

718. Shimizu, Y., Chou, A. N., Bando, H., Duyne, van, G., Clardy, J. C.: J. Am. Chem. Soc. *108*, 514 (1986).
719. Lund, A. E., Narahashi, T.: J. Pharmacol. Exper. Ther. *219*, 464 (1981).
720. Bercken, van den, J., Vijverberg, H. P. M.: Pestic. Chem. Human Welfare Environ. Proc. IUPAC Congr. Kyoto 1982 III, 115.
721. Vijverberg, H. P. M., Ruigt, G. S., Bercken, van den, J.: Pestic. biochem. Physiol. *18*, 315 (1982).
721a. Tippe, A.: Pestic. Biochem. Physiol. *28*, 67 (1987).
722. Vijverberg, H. M. P., de Weile, J. R., Ruight, G. S. F., Bercken, van den, J.: Abstr. Neurotox' *85*, 237, Symposion Bath April 1985.
723. Lund, A. L., Narahashi, T.: Neurotoxicology *3*, 11 (1982).
723a. Lund, A. L., Narahashi, T.: Pestic. Biochem. Physiol. *20*, 203 (1983).
724. Souyri, F.: Abstr. Neurotox' *85*, 231, Symposion Bath April 1985.
725. Chalmers, A. E., Miller, T. A., Ohlsen, R. W.: Abstr. Neurotox' *85*, 41, Symposion Bath April 1985.
726. DeWeille, J. R.: Diss. 1986 Univ. Utrecht.
727. Roche, M., Guillet, J. C.: Abstr. Neurotox' *85*, 115, Symposion Bath 1985.
728. Osborne, M. P.: Pyrethroid Insecticides Tables Ronde *37*, 5 (1980); Roussel Uclaf.
729. Narahashi, T.: Abstr. Neurotox' 85, 225, Symposium Bath, April 1985.
730. Leake, L. D.: Comp. Biochem. Physiol. C, *72*, 317 (1982).
731. Nicholson, R. H., Hard, R. J., Osborne, M. P.: Proc. Neurotox' *79*, 465 Symposion York Sept. 1979.
732. DeVries, D. H., Georghiu, G. P.: Pestic. Biochem. Physiol. *15*, 234 (1981).
733. Chialiang, Ch., Devonshire, A. L.: Pestic. Science *1982*, 156.
735. Gordon, M. A.: J. Immunopharmacol. *6*, 389 (1984).
736. Stelzer, K. J., Gordon, M. A.: Biochem. Biophys. Acta *812*, 361 (1985).
737. Bloomquis, J. R., Miller, T. A.: Abstr. Neurotox' *85*, 28, Symposion Bath April 1985.
738. Chang, C. P., Plapp, F. W.: Pestic. Biochem. Physiol. *20*, 86 (1983).
739. Gammon, D. W., Holden, J. S.: Proc. Neurotox' 79, 481, Symposion York Sept. 1979.
740. Fullbrook, S. L., Holden, J. S.: Proc. Neurotox' 79, 281 Symposion York, Sept. 1979.
741. Schnitzerling, H. J.: Pestic. Biochem. Physiol. *24*, 362 (1985).
742. Chang, C. P., Plapp, F. W.: J. Econ. Entomol. *76*, 1206 (1983).
743. Salgado, V. L., Irving, S. N., Miller, T. A.: Pestic. Biochem. Physiol. *20*, 100; 169 (1983).
744. Soderlund, D. M., Ghiasuddin, S. M., Helmuth, D. W.: Life Sci. *33*, 261 (1983).
745. Moser, R., Thomas, R. M., Gutte, B.: FEBS-Letters *1983*, 247.
746. Süddeutsche Zeitung 19.01.87.
747. Koch, R. B., Cutkomp, C. K., Do, F. M.: Life Science *8*, 289 (1969).
748. Desaiah, D., Cutkomb, L. K., Koch, R. B.: Pestic. Biochem. Physiol. *4*, 232 (1974).
749. Dary, C. C., Cutkomb, L. K.: Pestic. Sci. *1984*, 443.
750. Clark, J. M., Matsumura, F.: Pestic. Biochem. Physiol. *18*, 180 (1983).
751. Desaiah, D., Cutkomb, L. K.: Gen. Pharmacol. *6*, 31 (1975).
752. Matsumura, F.: Pestic. Chem. Human Welfare Environ. Proc. IUPAC Congr. Kyoto 1982 III, 3.
753. Schneider, R. P.: Biochem. Pharmacol. *24*, 939 (1975).
754. Rashatwar, S. S., Matsumura, F.: Pestic. Biochem. Physiol. *25*, 90 (1986).
755. Schouest, L. A., Salgado, V. L., Miller, T. A.: Pestic. Biochem. Physiol. *25*, 381 (1986).
755a. Luo, Y., Chang, T. T.: Kunchong Xuebao *30*, 8 (1987), C.A. *106*. 1710934.
        Chang, T. T., Wu, J. H., Luo, Y.: Abstr. Internat. Entomology Congr. Hamburg 1984, 16.1,2.
756. Matsumura, F.: Membrane Receptors and Enzymes as Targets of Insecticidal Action, Plenum Press 1986, p. 173.
757. Clark, M.: Diss. Abstr. Int. *42*, 2218 (1981).
758. Clark, J. M., Matsumura, F.: Pestic. Biochem. Physiol. *18*, 180 (1982).
759. Jones, O. T., Lee, A. G.: Pestic. Biochem. Physiol. *25*, 420 (1986).
760. Dary, C. C., Buessow, S. C., Wenzler, H., Cutkomb, L. K.: Abstr. Neurotox' 85, 51, Symposium Bath, April 1985.

761. Ghiasuddin, S. M., Kawauchi, S., Matsumura, F., Doherty, J. D.: Biochem. Pharmacol. *31*, 1483 (1982).
762. Eldefrawi, E. M.: Paper given at the Gordon Research Conf. Januar *1985*, Oxnard.
763. Sherby, M. S., Eldefrawi, A. T., Deshpande, S. S., Albouquerke, E. X., Eldefrawi, E. M.: Pestic. Biochem. Physiol. *26*, 107 (1986).
764. Rashatwar, S., Matsumura, F.: Biochem. Pharmacol. *34*, 1689 (1985).
765. Doherti, J. D., Lauter, C. J., Salem, N. J.: Comp. Biochem. Physiol. C.O. Comp. Pharmacol. Toxicol. *1986*, 373. .
766. Leake, L. D.: Abstr. Neurotox' *85*, 106 Symposion Bath April 1985.
767. Malenka, R. C., Madison, D. V., Nicholson, R. A.: Nature *321*, 175.
768. Kaczmarek, L. K.: TIBS *1987*, 30.
768a. Matsumura, T.: Lecture at the Gordon Res. Conf. Jan. 26th 1987. Sta. Barbara, Calif.
769. Bodnaryk, R. P.: Pestic. Biochem. Physiol. *18*, 334 (1982).
770. McKee, M. J., Knowles, C. O.: Econ. Entomol. *1984*, 1376.
771. Keyserlink, von, H. C.: Paper given at the International Entomology Congress, Hamburg Sept. 1984.
772. Ray, D. E.: Neurobehav. Toxicol. Teratol. *4*, 801 (1982).
773. Saad, A. S. A., Negem, S. S.: Meded. Fac. Landbouwwet Rijksuniv. Gent *47*, 679, 1982.
774. Katsuda, S., Hajima, H., Namite, N.: JA 602147604/84 C.A. 105, 43112s.
775. Satelle, D.: 190th ACS Meeting, Chikago Sept. 1985.
776. Narahashi, T.: 191th ACS-Meeting, April 1986 New York.
777. Abbassy, M. A., Eldefrawi, M. E., Eldefrawi, A. T.: Pestic. Biochem. Physiol. *19*, 299 (1983).
778. Abassy, M. A., Eldefrawi, M. E., Eldefrawi, A. T.: J. Toxicol. Environ. Health *12*, 575 (1983).
779. Bandyopadhyay, R.: Indian. J. Exp. Biol. *20*, 488.
780. Casida, J. E., Gammon, D. W., Glickman, A. M., Lawrence, L. J.: Ann. Rev. Pharmacol. Toxicol. *23*, 413 (1983).
781. Lawrence, L. J., Casida, J. E.: Science *21*, 1399 (1983).
782. Staatz, G. C., Bloom, A. F., Lech, J.: Toxicol. Appl. Pharmacol. *64*, 566 (1982).
783. Gammon, D., Casida, J. E.: Neurosci. Lett. *40*, 163 (1983).
784. Ozoe, Y.: Paper delivered at the Intern. Entomol. Conf. Hamburg Sept. 1984.
785. Lawrence, L. J.: Diss. Abstr. Int. B *45*, 843 (1984).
786. Matsumura, F.: Int. Entomol. Conf. Hamburg 1982, Abstr. R 16-1,3'.
787. Satelle, D.: Paper delivered at the Int. Entomol. Congr. Hamburg 1984.
788. Abalis, I. M., Eldefrawi, M. E., Eldefrawi, A. T.: J. Toxicol. Environ. Health *18*, 13 (1986).
789. Gammon, D. W.: Fundam. Appl. Toxicol. *5*, 9 (1985).
789a. Chalmers, A. E., Miller, T. A., Olsen, Z. W.: Pestic. Bioch. Physiol. *27*, 36 (1987).
790. Gammon, D. W., Sander, G.: Neurotoxicology *6*, 23 (1985).
790a. Sumitomo: JA 60214715.
790b. Eldefrawi, A., Abalis, I. M., Eldefrawi, M. E.: Membrane Receptors and Enzymes as Targets of Insecticidal Action, Plenum Press 1986, 107.
790c. Fisher, M. H., Mrozik, H.: Macrolide Antibiotics, Acad, Press 1984, XIV, 553.
791. Orchard, J.: Neurotox' *79*, 321, Symposion York Sept. 1979.
792. Norman, T. C.: Proc. Neurotox' *79*, 305, Symposion York Sept. 1979.
793. Soderlund, D. M.: Pestic. Biochem. Physiol. *12*, 38.
794. Sing, G. J. P.: Pestic. Biochem. Physiol. *25*, 264 (1986).
795. Orchard, J.: Pestic. Sci. *1983*, 229.
796. Dyball, R. E. J.: Pestic. Biochem. Physiol. *17*, 42 (1982).
797. Sing, G. J. P., Kundu, S. C.: Pestic. Biochem. Physiol. *18*, 158 (1982).
798. Chanh, P. H., Navarro-Delmasure, C., Chanh, A. P.: IRCS Met. Sci. Libr. Compend. *9*, 587 (1981).
798a. Brooks, M. W., Clark, J. M.: Pestic. Biochem. Physiol. *28*, 127 (1987).
799. Eells, J. T., Dubocovitch, M. L.: Abstr. Neurotox' *85*, 67, Symposion Bath April 1985.
800. Nicholson, R., Wilson, R., Potter, C., Bluck, M.: Pestic. Chem. Human Welfare Environ., Proc. IUPAC Congr. Kyoto 1982 III, 75.

800a. Kono, Y., Ozaki, N.: Appl. Entomol. Zool. *22*, 68 (1987).
801. Gusovsky, F., Hollingworth, E. B., Daly, J. D.: Proc. Natl. Acad. Sci. USA *83* (1986), Neurobiology 3003.
802. Stelzer, K. J., Gordon, M. A.: J. Immuno. Pharmacol. *1984*, 389.
803. Stelzer, K. J., Gordon, M. A.: Chem. Biol. Interact. *1985*, 105.
804. Jones, O. T., Lee, A. G.: Pestic. Biochem. Physiol. *25*, 420 (1986).
805. Stelzer, K. J., Gordon, M. A.: Biochem. Biophys. Acta *1985*, 812.
806. Stelzer, K. J., Gordon, M. A.: Pestic. Biochem. Physiol. *25*, 82 (1986).
807. Matsumura, F., Marshall, C. J.: Neurotoxicology *1985*, 271.
808. Wood McKenzie-Reports 1980 − March 1987.
809. Sateiz, D. E.: Informations Chimie *174*, 201 (1978).
810. Highwood, D. P.: Proc. Brit. Crop. Prot. Conf. *1979*, 361.
811. Jennings, A. G.: Pesticides *1985*, 23.
812. Farm Chemical Handbook 1982; Meister. Willoughby Ohio.
813. Farm Chemical Handbook 1986.
813a. Wood McKenzie-Report Jan. 1987.
814. Nachrichten für Außenhandel, Eschborn 2. 3. 87.
815. Black, I. A., Hewson, R. T.: Proc. Brit. Crop. Prot. Conf. *1979*, 377.
816. Winteringham, P.: Ann. Rev. Entomol. *14*, 409 (1969).
817. Ripcord Prospect Shell.
818. Breeve, M. H., Highwood, D. P.: Proc. Brit. Crop. Prot. Conf. *1977*, II, 641.
819. Fenvalerate Prospect, Sumitomo 1976.
820. C & E News Nov. 16th 1987.
821. Jacques, A. et al.: Anal. Methods Pestic. Plant. Growth Regul. *13*, 9 (1984).
822. TINS, *10*, 523 (1987).
823. Narahashi, T.: Paper given at Neurotox '88, Nottingham April 1988.
824. Tsushima, K., Yano, T., Takagaki, T., Matsuo, N., Hirano, M., Ohno, N.: Agric. Biol. Chem. *52*, 1323 (1988).
825. Matsumura, F.: Paper given at Neurotox '88, Nottingham April 1988.
826. Soderlund, D. M.: Paper given at Neurotox '88, Nottingham April 1988.
     Blomquist, J.R., ibid.
827. = 821.
828. Hellmuth, D. W., Ghiasuddin, S. M., Soderlund, D. M.: J. Agric. Food Chem. *31*, (1983) 1127.
829. Day, K., Kaushik, N. K.: Aquat. Toxicol. *10* (1987) 131.
830. Chang, J. T., Gao, P.: Insect Neurochem. Neurophysiol. [Pap. Int. Conf.] 2nd 1986, 239. Ed. A. B. Borkovec.
831. Takamatsu, Y., Kaneko, H., Akibo, J., Yoshitake, A., Miyamoto, J.: J. Pestic. Sci. *12* (1987) 397.
832. Tang, H. L., Ni, Y. S., Chang, J. T.: Kungchong Xuebao *30* (1987) 246.
     Tang, H. L., Ni, Y. S.: Insect Neurochem. Neurophysiol. [Pap. Int. Conf.] 2nd 1986, 409. Ed. A. B. Borkovec.
833. Sing, D. K., Agarwal, R. A.: Sci. Total Environ. *67* (1987), 263.
834. Tang, C., Ma, T., Liu, Y.: Huangjing Kexue Xuebao 7, (1987) 175.
835. Lagadic, L., Echaubard, M.: Poster presented at 9ème Colloque Physiologie de l'Insecte, 6–8 Sept. 1988; Lyon.
836. Saleh, M. A., Ibrahim, N. A., Soliman, N. Z., El Sheimy, M. K.: J. Agric. Food Chem. *34* (1986) 895.
837. County NATWEST Wood Mac Agrochem. Monitor 60, 9/88.
838. Kulzer, E., Fiedler, M., Kling, D., Kimmich, F.: Die Toxizität der Pyrethroide bei Süßwasserorganismen. By Ulmer, Stuttgart 1986.
839. Schnitzerling, H. J., Nolan, J., Hughes, S.: Experimental and Applied Acarology 6 (1989) 47−54
840. Martin, T. J., Hester, K. H.: Br, J. Dermatol. Syph. *53*, (1941) 127.
841. He, F., Wang, S., Liu, L., Chen, S., Zhang, Z., Sun, J.: Arch. Toxicol. *63*, (1989) 54.
842. Oortgiesen, M.: Dissertation Univ. of Utrecht 1989 p. 33

843. Broderick, M. P., Leake, L. D.: Progress and Prospects in Insect Controll p. 273 Proc. Intern. Conf. Reading 9/89, BCPC Monograph No. 43
844. Soderlund, D. M., Sanborn, J. R., Lee, P. W.: Progress in Pestic. Biochem. and Toxicol. III, 1983, p. 401 Ed. Hutson and Roberts.
845. Livingston, D. J., Ford, M. G., Buckley, D. S.: Neurotox '88 p. 469ff, p. 483ff
Molecular Basis of Drug and Pesticide Action, Internat. Symposium Series No. 832 Excerpta Medica.
846. Jørgensen, F. S., Byberg, J. R., Krogsgard-Larsen, P.: Neurotox '88, p. 497ff.
Molecular Basis of Drug and Pesticide Action, Internat. Symposium Series No. 832 Excerpta Medica.
847. Ruzo, L. O., Kimmel, E. C., Casida, J. E., J. Agric. Food Chem. 1986, 937

Subject Index

Previously Published Volumes in
Chemistry of Plant Protection

Chemistry of Plant Protection

Volume 5

K. Naumann, Leverkusen

Synthetic Pyrethroid Insecticides, Chemistry and Patents

1990. Approx. 350 pp. 5 figs.
ISBN 3-540-51314-0

This volume presents a detailed survey of the synthesis of pyrethroids and of almost every of the respective patent applications for synthetic pyrethroid active ingredients.

Contents: Synthesis of Pyrethroid Acids. – Synthesis of Important Pyrethroid Alcohols. – Formation of the Pyrethroid-ester-linkage. – Non-Ester Pyrethroids. – Analysis of International Pyrethroid Patent Activity.

Volumes 4 and 5 of the series provide the first complete and in-depth overview of synthetic pyrethroid insecticides and will be of utmost value to industrial chemists, biologists, molecular pharmacologists and patent professionals involved in pyrethroid chemistry.

Springer-Verlag Berlin Heidelberg New York London Paris Tokyo Hong Kong